Control of Energy Exchange

MODERN CONCEPTS IN MEDICAL PHYSIOLOGY

A MACMILLAN SERIES

Lysle H. Peterson, M.D., Consulting Editor

Loren D. Carlson, Ph.D., Ph.D. (Hon. Oslo)

PROFESSOR OF HUMAN PHYSIOLOGY, SCHOOL OF MEDICINE,
UNIVERSITY OF CALIFORNIA, DAVIS

Arnold C. L. Hsieh, M.D., D.Sc.

ASSOCIATE PROFESSOR OF HUMAN PHYSIOLOGY, SCHOOL OF MEDICINE,
UNIVERSITY OF CALIFORNIA, DAVIS

Control of Energy Exchange

The Macmillan Company | **Collier-Macmillan Limited,** London

First Printing, 1970

Library of Congress catalog card number: 72-103688

THE MACMILLAN COMPANY
866 THIRD AVENUE, NEW YORK, NEW YORK 10022
COLLIER-MACMILLAN CANADA, LTD., TORONTO, ONTARIO

Printed in the United States of America

Preface

THIS MONOGRAPH CONSIDERS ENERGY EXCHANGE in terms of the various systems that are involved. The first chapter presents general considerations of energy balance and includes discussions of normal weight, body composition, and control of food intake. The specific elements in the balance are dealt with in subsequent chapters on metabolism (heat production and the endocrine control of metabolism), expenditure of energy (work), and thermophysiology (physical factors). Specific considerations of temperature regulation are discussed in a separate chapter, which also includes the concept of normal temperature and its variations as well as the basic factors in the regulatory system—sensory input, central integrating system, and effector systems. Alterations in the homeostatic mechanisms caused by exposure to heat and cold are considered in a final chapter on adaptation. For the reader who requires more detailed exposition on any of these subjects, general and specific references are listed at the end of each chapter.

A standard set of symbols is considered essential and, therefore, is included in Appendix I. Also incorporated in the back matter of the book are a set of conversion tables and discussions of basic principles in the measurement of temperature, energy exchange, and body composition.

Control of Energy Exchange is neither a definitive reference nor a handbook. Rather, this brief volume has been written as an introduction to the subject for medical students, graduate students, and engineers interested in biologic phenomena. It should serve the function of consolidating pertinent information (generally scattered throughout the larger, standard textbooks of physiology and biochemistry) and making it available in a more compact and convenient format.

The authors wish to express their thanks and gratitude to Mrs. Sharon Holley, who assisted with the preparation of the manuscript. They also acknowledge with appreciation the many authors and publishers who gave them permission to reproduce various figures and tables. Complete credit to these original sources of publication is given at appropriate intervals throughout the text.

L. D. C.
A. C. L. H.

Contents

CHAPTER

1

Energy Balance

A SIMPLE ENGINE consumes a fuel to do work. The energy that does not appear as work is dissipated as heat. The body may be considered as a machine that does its own maintenance and powers a convective system that distributes fuel and oxygen to the local machine sites and removes the waste products and heat. The machine has an optimum operating temperature. In addition to internal maintenance, regulation, and computation, the body may move and do work. This monograph considers the machinery of the body involved in energy exchange—an exquisite system which maintains the composition of the machine and its temperature.

Energy In = Energy Out

The overall equation for the exchange of energy between the animal and its environment is $E_{in} = E_{out}$, but the concept of equality has meaning only for values integrated in long periods of time. Over shorter periods, the exchange may depart significantly from equality. Stores are used and deficits replenished. It is, therefore, convenient to treat the study of energy balance as an exercise in accountancy. If E_{in} is considered as the caloric fuel intake, a lack of balance with E_{out} as heat or work will appear in the accounts as a change in body weight. If E_{in} is considered as heat input from combustion of the fuels, a lack of balance with heat loss will appear as a change in body temperature. It is possible to have simultaneous changes in body weight and body temperature. Figure 1-1 shows the principal pathways from E_{in} to E_{out}.

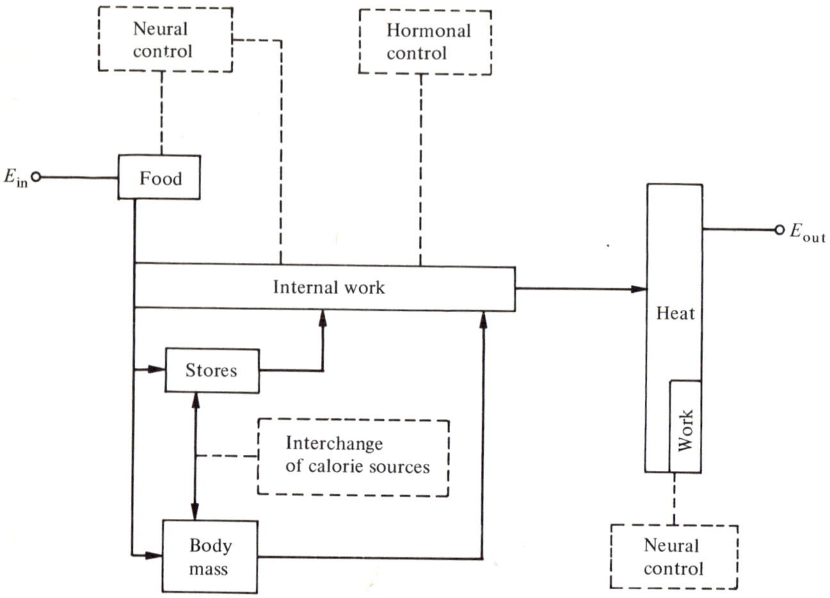

FIGURE 1-1: The caloric model showing the principal paths from energy intake (E_{in}) to energy output (E_{out}). The control loop is completed by Figure 1-8.

Energy enters the system as food with a caloric value dependent on its chemical nature. Ingestion with respect to time and caloric amount is regulated by the central nervous system. Ingestion and digestion of food require internal work and the food may appear in storage as carbohydrate or fat or as a part of the lean body mass, protein. The food may be oxidized to provide heat or work as the energy output of this system. The regulative processes of internal work are controlled by the endocrine and nervous systems. The temperature of the system is maintained by a separate controlling function, as is the physical work to be done.

Calorie Intake

There are two major phases in the consideration of the control of energy or food intake. *In the adult* the needs for internal maintenance, work, and temperature regulation force the mechanisms of balance in the system. *In young animals* growth plays an additional and important part in determining calorie intake. In the adult, weight is ordinarily kept constant but there are notable exceptions, such as the increase in body mass in the human during training or the decrease during bedrest or the continuous growth in certain animals such as the male rat.

Efficiency of Growth

It is a matter of experience that patterns of growth are uniform in animals and man. The growth rate and energy utilization for body construction follow a characteristic pattern for animals of a variety of species. The gross efficiency of growth (conversion of food to body weight) can be expressed as a simple logarithmic function of time.

$$E_t = E_0 - a \log t$$

where E_t = efficiency of conversion of food to body weight at time t,
E_0 = efficiency of conversion of food to body weight at 0 time, and,
a = the constant representing the decrease in efficiency with time.

The efficiency at 0 time (weaning) is 0.81 in a male rat. The over-all initial efficiency of food utilization at the beginning of postnatal growth is 0.35. This appears to be independent of size.

Values for the decrease in efficiency a have not been determined in a variety of species, including man. However, at maturity body synthesis does not cease but the net difference becomes 0; that is, ($E_0 - a \log t$) is 0. Thus, a could be computed if E_0 and the time to reach maturity were known. The gross efficiency may be of value for assessment of nutritional deficiencies.

An estimate of E_t can be obtained experimentally by first restricting food consumption to the amount necessary for weight maintenance and then noting the increase in weight and food consumption on resumption of *ad libitum* feeding. Kaunitz and coworkers (5)* used this method with rats in a warm environment and obtained a value of 1 gm of food for 1 gm of body substance formed. Hsieh and Ti (4), plotting the change in body weight against food intake of rats exposed to cold, obtained a value of 0.516 (Figure 1-2).

The mechanisms of control of synthesis of body tissues are not known. Several hormones, such as somatotrophin, thyroxin, and corticosterone, are essential for the process.

When the over-all efficiency E_t is 0, the body maintains its weight. There is a continual turnover of body mass; some cells of the body have relatively short half-lives (intestinal mucosa, two days); others are irreplaceable as cells (central nervous system), although these cells do degenerate.

At equilibrium, animals maintain weight by ingestion of the appropriate amounts of calories. Adolph (1) tested this assumption in rats by feeding them solid food diluted with cellulose or kaolin or a liquid diet diluted with water. The rats increased the volume of food ingested as the food was diluted. Adolph concluded that, within limits, rats eat for calories, although when fed highly concentrated foods the rats tended to gain weight.

*Parenthetical numbers pertain to the References at the end of the chapter.

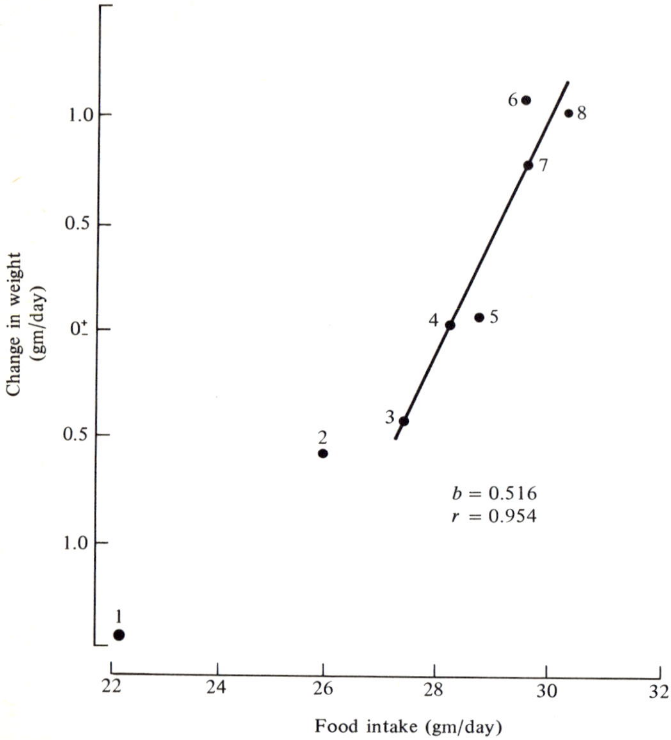

FIGURE 1-2. The relationship between change in weight and food intake of rats during exposure to a room temperature of 4°C. The number near the points indicates the number of weeks of exposure to cold. During the fourth and fifth weeks, there was no change in body weight, and it may be assumed that 28 gm/day is required for maintenance of body weight at 4°C. The slope *b* indicates the change in weight per gram of food intake. (From A. C. L. Hsieh, and K. W. Ti. *J. Nutr.* **72:** 283–288, 1960.)

Effect of Energy Requirement

If the requirements for energy are increased, animals will increase their food intake. Rats living at 5°C will consume twice as much food as their litter mates living at 25°C. When animals are exercised, they increase their food consumption. Mayer (7) exercised rats for varying periods during the day up to the point of exhaustion. Food intake increased regularly (Figure 1-3) with exercise, with two interesting deviations in the function. At low levels of activity body weight tends to increase, and at high levels of activity food consumption does not meet the demand and body weight decreases.

The preceding experiments suggest that the level of energy output (activity) is a factor regulating the level of energy intake. There is also evidence for a reverse effect, that is, the level of energy intake acts as a limiting factor for the level of activity. This relationship was demonstrated in humans during World War II by Kraut and Müller (6), who calculated the energy balance in workers on rations and showed that an inadequate intake will result in a decrease in work output. They also demonstrated the value of incentives in increasing productivity. The results could have been predicted on theoretic grounds, but the observations demonstrate, in practical terms, that rationing of food leads to rationing of industrial production. Thus, in times when availability of food is controlled by a rigid food rationing system, control of industrial production is diverted from management to those responsible for food administration.

An interesting relationship between energy intake, body weight, and activity in an industrial male population in West Bengal has been reported by Mayer and coworkers (8) (Figure 1-4). Calorie intake was estimated from dietary interviews and work output characterized by profession. There is a gradual reduction in energy intake and body weight with increasing levels of activity from stallholders to mechanics, suggesting that socioeconomic factors may be involved. In the range of activity, mechanics to carriers, body

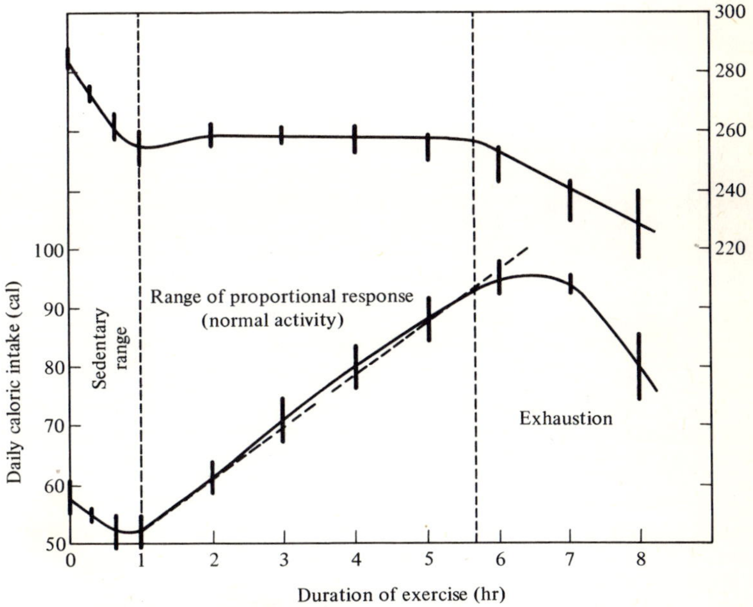

FIGURE 1-3. Daily calorie intake of rats exercised for varying periods of time during each day. Body weight is given in the upper section. Bars denote standard deviation. Note that small amounts of exercise (sedentary range) tend to reduce weight and food intake, whereas body weight is maintained over a wide range of exercise (1 to 5 or 6 hr) and food consumption nearly doubles. Exhausting work decreases food consumption, resulting in weight loss. (From J. Mayer, et al. *Amer. J. Physiol.* **177:** 544–548, 1954.)

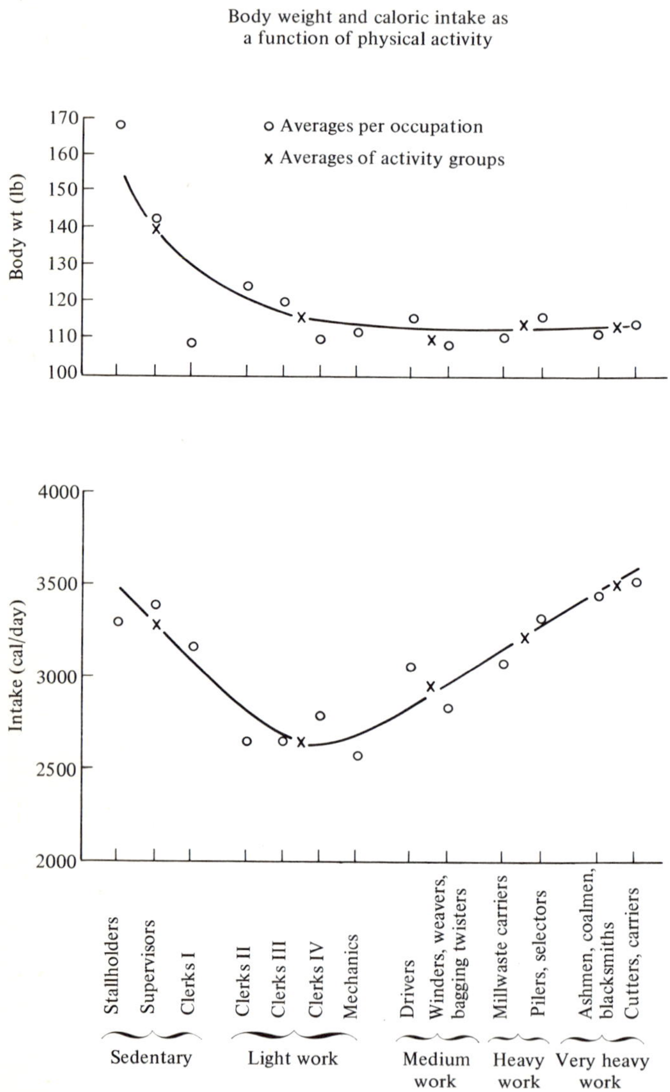

FIGURE 1-4. The relationship of calorie intake and body weight to activity in adult men. The individuals are grouped according to the type of work they perform and the groups arranged in ascending order of physical activity. Note that body weight decreases as the activity increases from sedentary to light work and then remains constant. The increased body weight at low levels of activity probably indicates an increase in body fat. The increase in food consumption of the sedentary group may be attributed to affluence or boredom. (From J. Mayer, P. Roy, and K. P. Mitra. *Amer. J. Clin. Nutr.* **4:** 169–175, 1956.)

weight was constant but energy intake increased with increases in the degree of activity. The fact that body weight remained constant leads one to conclude that there is a considerable difference in body composition between the groups.

The Concept of Normal Weight

Maintenance of "normal" weight has many connotations. It implies that there are sequelae related to being underweight or overweight. Indeed, there are tables of average weights related to height composed in the usual statistical manner and tables of desirable weights in adults which are the weights of insured persons with the lowest mortality.

The most common cause of being underweight is malnutrition. Undernourishment or lack of specific dietary components lead to a loss in weight. Certain clinical conditions such as hyperthyroidism are also accompanied by loss of weight.

Two major syndromes of protein-calorie malnutrition of early childhood are kwashiorkor and nutritional marasmus. Kwashiorkor is almost never exclusively dietary in etiology. But it is of interest that a similar syndrome can be produced experimentally by feeding a low protein, mainly carbohydrate diet to piglets. Nutritional marasmus is principally due to a diet that is low in both protein and calories (so-called balanced starvation). It occurs most commonly in the first year of life as a result of a failure of lactation and of attempted artificial feeding with very dilute milk.

When intake fails to match output, a utilization of stores must occur. Because the intake of food is periodic and energy expenditure is continuous, though variable, stores are always being used and then restored. The body uses carbohydrate and fat for energy. When an individual fasts or is starved, carbohydrate stores are depleted rapidly; fat and then body protein serve as the major long-term sources of energy. The depletion of stores over a period of time is shown in Figure 1-5. When the fasting individual exercises, the energy provision from the different stores are time-dependent also (Figure 1-6). Although still ill-defined experimentally, there is evidence that the utilization of stores is an adaptive function.

The most common cause of being overweight (obesity) is excessive eating, although this diagnosis should never be made without other clinical tests. Obesity is a problem of an affluent society and occurs in those geographic areas and those periods of history characterized by abundance.* Obesity is associated with a number of clinical disorders, such as coronary disease, and reduced longevity. Although obesity is obviously the result of an imbalance of caloric intake and output, this fact does not give the cause. In addition to psychologic factors associated with appetite and hunger, the following pages

*Cultural and sociologic factors are contributory to the problem. In many societies obesity is regarded as a desirable attribute.

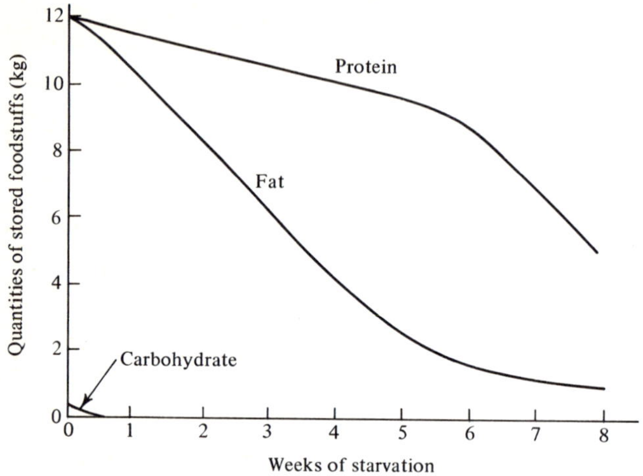

FIGURE 1-5. The reduction in food stores during starvation. The carbohydrate stores last only a few hours, lipid serves as the major source of energy for a period of weeks, and protein acts as a final reserve. (From A. C. Guyton. *Textbook of Medical Physiology;* 3rd ed. Philadelphia: Saunders; 1966.)

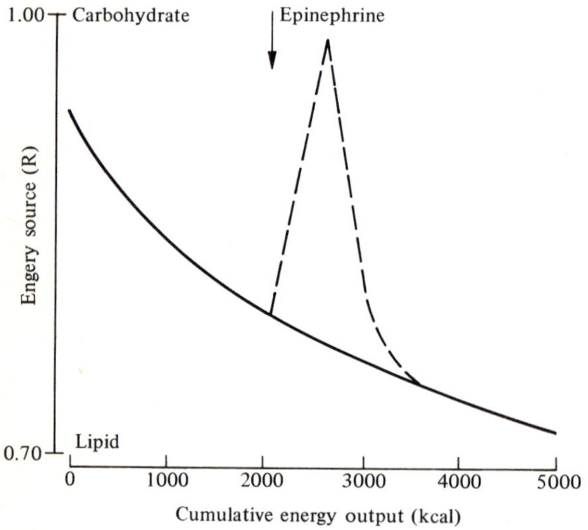

FIGURE 1-6. Relative carbohydrate and lipid utilization of a fasting man during 24 hr in which he expended a total of 5,090 kcal of energy by alternate periods of hard work (O_2 consumption 1.75 L/min) and rest. The administration of epinephrine as indicated by the dashed line temporarily caused an increase in carbohydrate utilization. (Adapted from D. B. Dill, H. T. Edwards, and R. H. deMeio, *Amer. J. Physiol.* **111:** 9–20, 1935.)

will suggest some possible neural, endocrine, and biochemical explanations for the syndrome. Careful biochemical characterization of the obese person has led to a variety of theories concerning obesity. The etiology still remains complex.

Body Composition

In addition to the establishment of a normal weight, there is a normal body composition, which is sex dependent. The division in body compartments is given in Table 1-1. On a relative basis a normal male has one-half the lipid content of a normal female. A thin man may have twice the total body water of an obese woman. Were these individuals of the same weight, a substance introduced into the water compartment of the body would have twice the final concentration in the female.

It has been established that the fraction of nitrogen and the fraction of water in the fat-free body mass are constants. Whole body water on a fat-free basis is 0.732 times the fat-free body mass (also termed *lean body mass*). With such uniformity in the composition of the lean body mass, it is possible to define the relationship between total body water and the fraction of total body fat as

$$\text{Fraction of fat} = 1.000 - \frac{\text{fraction of water}}{0.732}$$

In the male of normal or average weight the fraction of body water is 0.60, and in the female of average weight the fraction is 0.50. (See also Table A-10 of Appendix III.)

TABLE 1-1
The Approximate Fractions of the Fat, Fat-free, and Water Compartments of the Body

		Fat	Fat-free Substances	Water
Male	Average weight	0.18	0.22	0.60
	Thin	0.04	0.26	0.70
	Obese	0.32	0.18	0.50
Female	Average weight	0.32	0.18	0.50
	Thin	0.18	0.22	0.60
	Obese	0.42	0.16	0.42

Specific Gravity

The fraction of fat in the body determines its specific gravity. Body fat may be calculated if the body's specific gravity is known. An empirical formula derived is

$$\text{Fraction of fat} = \frac{5.548}{\text{sp. gr.}} - 5.044$$

The specific gravity may be measured by water displacement or weighing underwater. The fraction of fat may be estimated by measurement of skinfolds. (See also Table A-10 of Appendix III.)

Control of Food Intake

Brobeck (2) and his colleagues postulate that the day-to-day regulation of food intake is in terms of not a definite quantity of energy but the specific dynamic action of the food. This hypothesis is supported by evidence that food consumption is related to environmental temperature and that warming the hypothalamus inhibits feeding.

Jean Mayer has suggested a glucostatic mechanism. Using data such as those shown in Figure 1-7, Mayer proposed that the central nervous system maintains, in the lateral hypothalamus, "glucoreceptors" sensitive to fluctuations of available blood glucose and responsible for the control of appetite. The hypothesis was derived from measurements of the glucose concentration difference between finger blood and antecubital vein blood. When the glucose concentration difference is small, hunger ensues; when the difference is large, satiety results.

The fundamental hunger "drive" initiates a complex behavior pattern. The relationships between food and feeding and between appetite and satiety involve a number of interacting factors and reflex pathways.

Part of the difficulty in the acceptance of any single hypothesis concerning food intake is the multiplexity of factors which influence this activity. Some of these are shown in Figure 1-8. The primary pathway involves the drive of hunger and the search for food. Selection and ingestion are complex behavioral patterns. Certain areas of the brain stem are essential for the reflection of the effects of temperature and blood glucose concentrations as well as the influence of feeding states such as boredom and visceral sensations from the stomach and the effects of certain drugs. Hunger, the awareness of the need to ingest food, is the basic primary drive. Appetite is the desire to ingest food and is more likely to be related to taste and appeal of food than to energy requirements. Satiety and anorexia describe two entirely different states of lack of desire to eat—the former and more pleasant state occurs after the

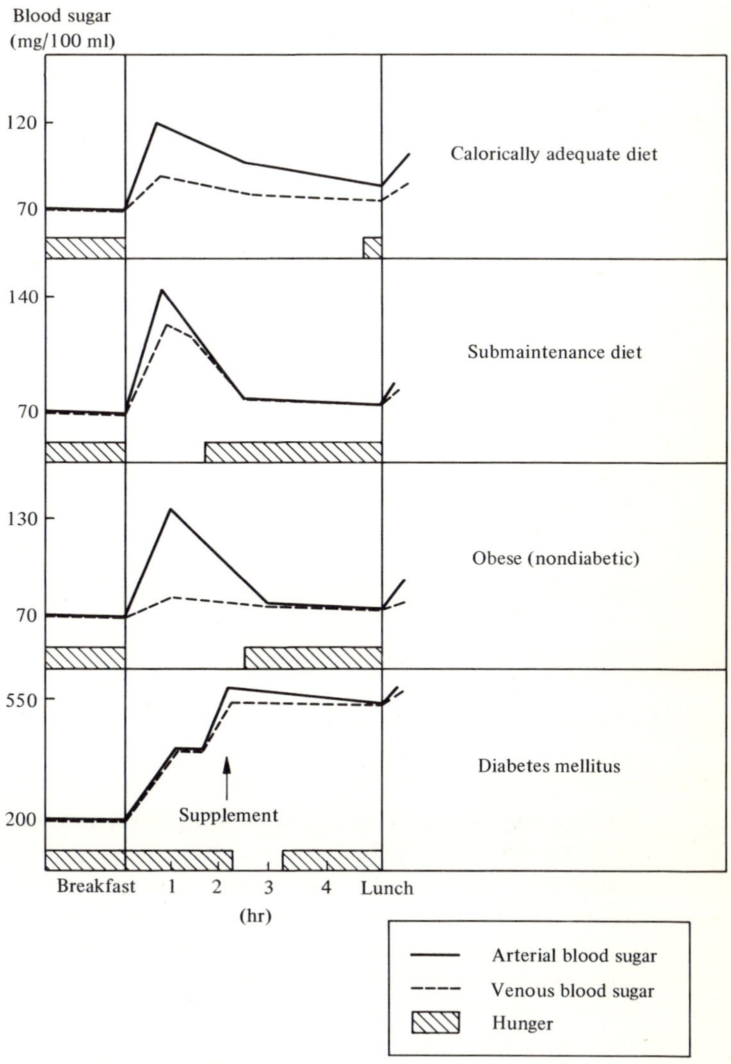

FIGURE 1-7. Arterial and venous blood glucose concentration and hunger feeling recorded during the morning. The size of the peripheral glucose arteriovenous difference correlates negatively with hunger feelings. No hunger feeling appears if the difference is greater than 10 mg per 100 ml. (From J. Mayer. In: *Fat Metabolism*. Edited by V. A. Najjar. Baltimore: Johns Hopkins, 1954.)

ingestion of food; the latter occurs in spite of the physiologic requirement for food.

Central control of food intake is primarily found in hypothalamic regions, although clearly the basic patterns of feeding involve reflex pathways requiring integrated motor activity and a variety of motor and sensory components in

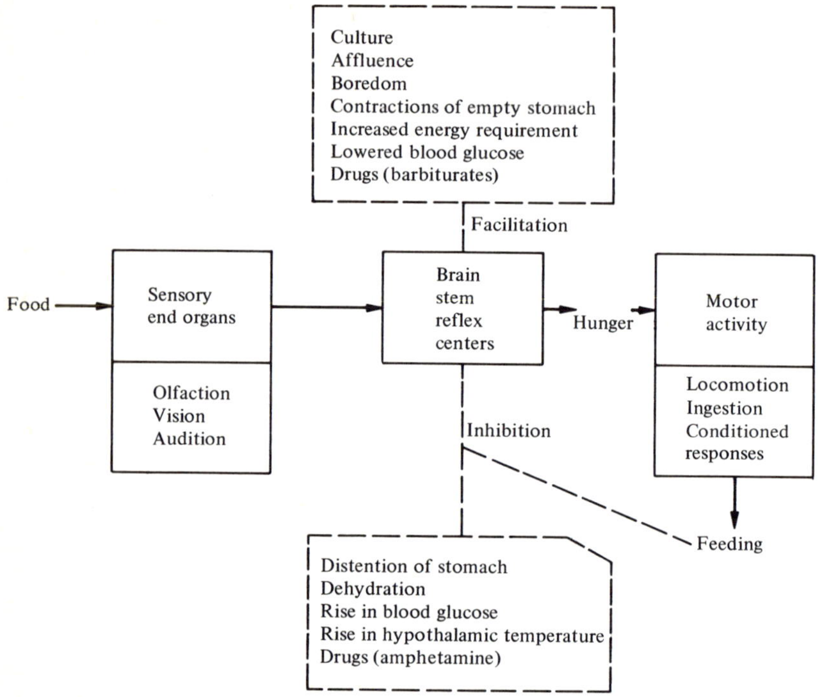

FIGURE 1-8. Pathways concerned with feeding.

the nervous system. Two major areas in the hypothalamus have been found to be related to feeding. The medial hypothalamus has been demonstrated to function in reactions of satiety, and the lateral portions of the hypothalamus are responsible for hunger, or the drive for feeding. Destruction of the lateral regions causes animals to fail to eat, whereas medial lesions lead to overeating and obesity. Stimulation of these respective areas also confirms the functional pattern described. When the lateral hypothalamus is stimulated in unanesthetized animals, feeding occurs, whereas stimulation to the medial region causes the cessation of feeding. Whether the hypothalamus acts as an appestat or more as a calorimetric satiety center is still a matter of debate, and there are a number of hypotheses concerning the mechanism of actions in the hypothalamus.

The part played by the endocrine system in the control of food intake is by no means clear. In Simmonds' disease (an extreme form of absolute panhypopituitarism most commonly caused by postpartum complete necrosis of the pituitary gland) cachexia may be severe. However, intractable loss of appetite is not equally frequent in all types of panhypopituitarism. Hypothyroidism results in a reduced food intake. In hyperthyroidism food intake is increased but is not adequate to maintain weight, and the severe hyperthyroid may literally consume his body protein. When animals are thyroidectomized, food consumption decreases and body weight falls. The time course for the weight loss closely parallels the reduction in metabolism.

References

1. Adolph, E. Urges to eat and drink in rats. *Amer. J. Physiol.* **151:** 110–125, 1947.
2. Brobeck, J. Food and temperature. *Recent Progr. Hormone Res.* **16:** 439–459, 1960.
3. Dill, D. B., H. T. Edwards, and R. H. deMeio. Effects of adrenalin injection in moderate work. *Amer. J. Physiol.* **111:** 9–20, 1935.
4. Hsieh, A. C. L., and K. W. Ti. The effects of L-thyroxine and cold exposure on the amount of food consumed and absorbed by male albino rats. *J. Nutr.* **72:** 283–288, 1960.
5. Kaunitz, H., C. A. Slanetz, and R. E. Johnson. Utilization of food for weight maintenance and growth, *J. Nutr.* **62:** 551–559, 1957.
6. Kraut, H. A., and E. A. Muller. Caloric intake and industrial output. *Science* **104:** 495–497, 1946.
7. Mayer, J., N. B. Marshall, J. J. Vitale, J. H. Christensen, M. B. Mashaykhi, and F. J. Stare. Exercise, food intake and body weight in normal rats and genetically obese adult mice. *Amer. J. Physiol.* **177:** 544–548, 1954.
8. Mayer, J., P. Roy, and K. P. Mitra. Relation between caloric intake, body weight and physical work: Studies in an industrial male population in West Bengal. *Amer. J. Clin. Nutr.* **4:** 169–175, 1956.

Additional Reading

Anand, B. K. Central chemosensitive mechanisms related to feeding. In: *Handbook of Physiology. Sec. 6. Alimentary Canal, Vol. I.* Edited by C. F. Code. Washington: American Physiological Society, 1967, pp. 249–263.

DeGroot, J. Organization of hypothalamic feeding mechanisms. In: *Handbook of Physiology. Sec. 6. Alimentary Canal. Vol. I.* Edited by C. F. Code. Washington: American Physiological Society, 1967, pp. 239–247.

Kleiber, M. *The Fire of Life.* New York: Wiley, 1961.

Yamamoto, W. S., and J. R. Brobeck, eds. *Physiological Controls and Regulations.* Philadelphia: Saunders, 1965.

CHAPTER

2

Metabolism

The Metabolic Rate

ENERGY BALANCE MODEL

To explore the questions concerned with energy balance, the system may be analyzed in various configurations. The most general is given in Figure 2-1. The energy that may be derived from food by oxidation is expressed in kilocalories per gram. The values may be determined by conventional bomb calorimetry, calculated from bond energies, or empirically by a combination of direct and indirect total body calorimetry. Three principal categories of foodstuffs serve as sources of energy. The significant values for these are given in Table 2-1.

Energy has many units, the common unit in physiology is the kilocalorie. The rate of energy conversion per unit time, power, is usually expressed as kcal/hr. This can be converted to the recommended standard physical unit, joules per second, or watts, by multiplying by 1.16.

Energy out in the simple system will appear as work or heat. The usual expression as work is in kg • m/min (1 kilocalorie/min = 426.85 kilogram meters/min). The efficiency of the body in work is usually less than 20 per cent.

Terms of balance are in the units of weight or heat as equivalents. If the energy coming into a system is not in balance with the energy output, a change in weight will occur—food will either be stored or removed from stores. With respect to heat, a temperature equilibrium is maintained and imbalances must give rise to changes in temperature.

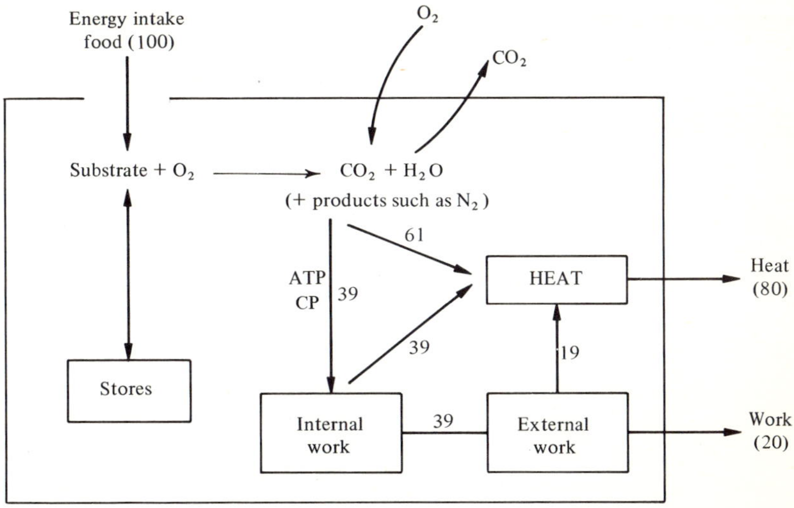

FIGURE 2-1. The energy balance model. Potential energy (100) enters the system as food and is released by oxidation. Part of the energy (39) is stored as adenosine triphosphate (ATP) and creatine phosphate (CP), which act as immediate sources of energy for work, part (61) is released as heat. During internal or external work, more heat is released. The efficiency of the body in work is about 20 per cent.

CALORIMETRY

Direct calorimetry measures heat output over a period of time. The basic assumption is that the subject is in temperature equilibrium. Techniques of direct calorimetry are given in Appendix III.

Indirect calorimetric measures derive heat production from gaseous exchange from the basic data given in Table 2-1. Because there are three combustible substances, the determination of amounts of each of the three cannot be determined from the gaseous exchange. The amount of protein burned is calculated from the urinary nitrogen. Subtracting the oxygen equivalents permits the calculation of a nonprotein respiratory quotient *R*, from

TABLE 2-1
Calorie Value and Gaseous Exchange in the Oxidation of Categories of Food

Food	kcal/gm	LO_2/gm	LCO_2/gm	kcal/LO_2	kcal/LCO_2	R
Carbohydrate	4.1	0.81	0.81	5.05	5.05	1.00
Lipid	9.3	1.96	1.39	4.74	6.67	0.71
Protein*	4.2	0.94	0.75	4.46	4.57	0.80

* Protein oxidation exclusive of N or S moiety. One gm of urinary nitrogen is equivalent to 6.25 gm of protein oxidized.

which the relative amounts of fat and carbohydrate can be determined. Several techniques for determination of gaseous exchange and the conversion method are given in Appendix III.

BASAL METABOLIC RATE

The reference value for the metabolic rate is the basal metabolism. This measurement is made under a controlled set of circumstances. The variables controlled are the following:

1. *Activity.* The measurement is made at rest, at least 30 min after activity and 12 hr after heavy exercise.
2. *Nutrition.* The measurement is made in the postabsorptive state—12 hr after food intake. The principal variable here is the specific dynamic action of food. Because the major effect is from protein, additional qualifications concerning meals prior to the test may be required.
3. *Environment.* A thermally neutral environment is necessary to ensure against metabolic adjustments in temperature regulation. Emotion-provoking sounds and conversations must be avoided.

To provide a reference value, a number of variables must be taken into account:

1. *Body Size.* In the human the metabolic rate is usually expressed in relation to body surface (square meters). This value has been empirically found to be related to the height and weight of the individual (S.A. $[m^2] = Ht^{0.725}cm \times Wt^{0.425}kg \times 0.00718$). A table for this conversion is given in Appendix III, Table A-12. In animals (and perhaps man should be included) the 3/4 power of the body weight is recommended as representative of metabolic body size. The metabolic rate would be expressed as $kcal/kg^{3/4}$ hr. A conversion table is given in Table A-13 of Appendix III.
2. *Age.* The growing animal is not in thermodynamic equilibrium because food is being converted to increase mass. In the adult human a correction is made for age. This correction has been empirically determined from a group of persons judged to be clinically normal with respect to metabolic disorders.
3. *Sex.* In the human a small sex difference exists in metabolic rate in relation to surface area.
4. *Temperature.* Metabolism and body temperature are dependent. Each 1°C change alters metabolism by about 13 per cent.

Basal metabolism is markedly altered by thyroid disease. The data plotted in Figure 2-2 indicate the distribution of BMR expressed as:

$$\frac{\text{Measured basal metabolism} - \text{standard}}{\text{Standard}} \times 100$$

This distribution plot is indicative of the range of normal individuals, individuals in the nonmetabolic diseases group, and the extent of the deviation in thyroid disease (4).

The basal metabolism is a combination of cellular processes related to metabolic turnover of constituents and maintenance of cell composition in addition to the work of certain systems such as the central nervous system, heart, lungs, gastrointestinal tract, and kidney. Assuming that the systems supporting the metabolic mass would be proportional to that mass, physiologists have attempted to define the metabolic mass. However, different tissues have different metabolic rates. For example, the brain, which is 2 per cent of the body weight, contributes 16 per cent of the body heat production; skin and muscle, which comprise 52 per cent of the body mass, represent 25 per cent of the resting heat production. Other interesting interrelations with respect to energy exchange are illustrated in Figure 2-3. Nominally, the metabolic rate in man in the basal state is 40 kcal/hr$\cdot$m^2 or 68 kcal/hr for a 70-kg man. This is 1 kcal/kg or 2.8 kcal/kg$^{3/4}$. In brain this relationship is 8 kcal/kg$^{3/4}$, and in resting muscle it is 0.86 kcal/kg$^{3/4}$ (Figure 2-3).

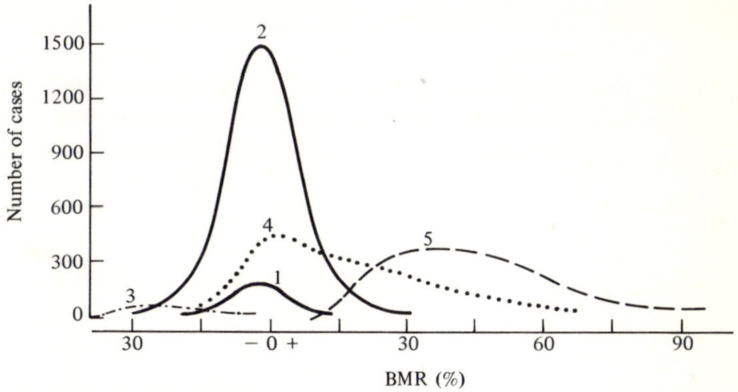

FIGURE 2-2. Comparison of (1) normal distribution of basal metabolism with those of groups having (2) a nonmetabolic disease (3) a spontaneous myxedema, (4) an adenomatous goiter with and without clinical hyperthyroidism, and (5) an exophthalmic goiter. (Adapted from W. M. Boothby, J. Berkson, and W. A. Plummer, *Ann. Intern. Med.* **11:** 1014–1023, 1937.)

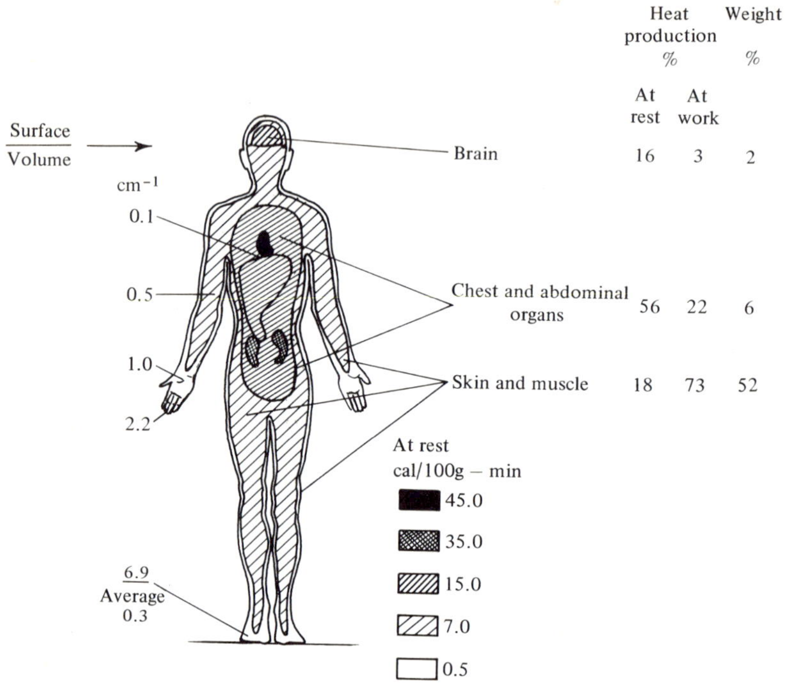

FIGURE 2-3. Heat production at rest and in work of various parts of the body. (Adapted from J. Aschoff and R. Wever. *Die Naturwissenschaften* **45:** 477, 1958.)

Regulation of Metabolism

The naive notion that basal metabolism might be regulated by (albeit proportional to) the surface area or lean body mass is made untenable by the marked discrepancies in the level of metabolism of different organs. The ultimate control of metabolism lies at the subcellular level and the mechanisms will be found in the rate-limiting factors in cell respiration (6).

RATE-LIMITING FACTORS

Three categorical divisions can be made of the control mechanisms that limit cell respiration and function.

1. Substrate diffusion or transport possibly related to structure of the cell membrane. At the macromolecule level, binding sites of reactive groups or their proper sequential appearance may exert controls.
2. Mass action criteria concerned with reactant concentration.
3. Control of reaction rates by catalysts (enzymes), cofactors, and essential ions. Although the principles of control will be developed as general

for all cells, different cells have a highly specific reactivity to the surrounding media.

Mass action criteria and kinetic considerations relate to the principal pathways of energy production of a cell and the alternate pathways, which appear in character as shunts, cycles, or cascades. A multipathway system links the substrate, such as glucose, with oxygen. About 39 per cent of the resulting energy change in the system is storable in specific chemical bonding and will eventually reappear as heat. Because of inefficiency of the system, about 61 per cent of the potential energy appears directly as heat. The net function is system maintenance. The system has three major steps (Figure 2-4): the phosphorylation of glucose, the glycolytic phosphorylation of ADP, the oxidative phosphorylation of ADP. Alternate pathways provide for the oxidation of lipid as a substrate and its reversible transfer to glucose and glycogen and the oxidation of deaminated amino acids and their conversion to glucose or glycogen.

A lowered blood sugar or a change in membrane permeability to glucose might limit metabolism because of substrate depletion or the required use of other substrates to meet the demand. Substances such as thiamine (vitamin B_1) are necessary as indispensable coenzymes (cocarboxylase). Certain hormones may act by influencing the activity of adenyl-cyclase, which, in the

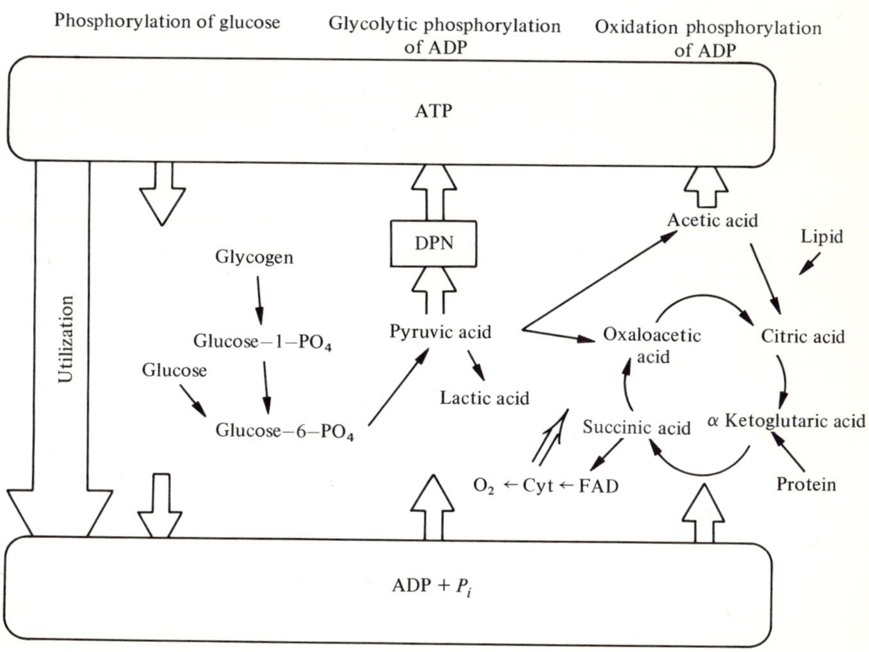

FIGURE 2-4. The multipathway system linking substrate with oxygen.

presence of Mg^{++} and Ca^{++} ions, converts ATP to cyclic-3′5′-AMP, a substance involved in the formation of active phosphorylase. An excess of ADP without complete depletion of ATP would stimulate metabolism by addition to the ADP pool.

An analogue of feedback regulation in biochemical systems has been developed in detail by Britton Chance (5). The chemical analogue of an electrical feedback amplifier sets voltage as an analogue of chemical concentration. Feedback may be provided by concentrations of intermediates or the output chemical. Multipliers are reaction rate constants. A chemical "feedback" is obtained by integration of the rates.

REGULATION OF ENERGY UTILIZATION

One of the intriguing elements in the regulation of energy exchange is the mechanism that establishes body composition—primarily the deposition of lipid in the adult. Bone structure and muscle mass are related to use, whereas lipid stores appear to be inversely related to the general activity of the animal and show sex, age, ethnic, and genetic differences. Lipid serves as the long-term energy storage pool of the body (the bonds or investment account); glycogen is more short-lived (the bank account); and bond energies, such as those in ATP and creatine phosphate, serve as immediate energy sources (the ready cash).

Skeletal, muscular, and organ development during growth are intricately involved with endocrine factors and proper nutrition. The prenatal animal has little stored lipid and is nearly 95 per cent water. As it grows it rapidly deposits lipid in adipose tissue. The regulation of stores and the basal level of expenditure are involuntary and involve autonomic and endocrine control.

Metabolic expenditure is under voluntary control insofar as activity is concerned. Depletion of readily accessible stores (ATP) and the increase in phosphate acceptor (ADP) will be dependent on the action elicited by motor activity—ordinarily smooth or striated muscle contraction.

Metabolic Endocrines

INSULIN AND GLUCAGON

Insulin is a hormone formed in the pancreas; it decreases liver and muscle glycogenolysis and facilitates glucose transfer into cells. Thus blood glucose is inversely related to insulin level. Glucagon, also a pancreatic hormone, increases liver glycogenolysis. Figure 2-5 depicts the feedback mechanisms regulating the secretion of these two hormones. When blood glucose concentration rises (indicated by the upward-pointing arrow inside the box labeled blood glucose), pancreatic secretion of insulin is stimulated. Depression of

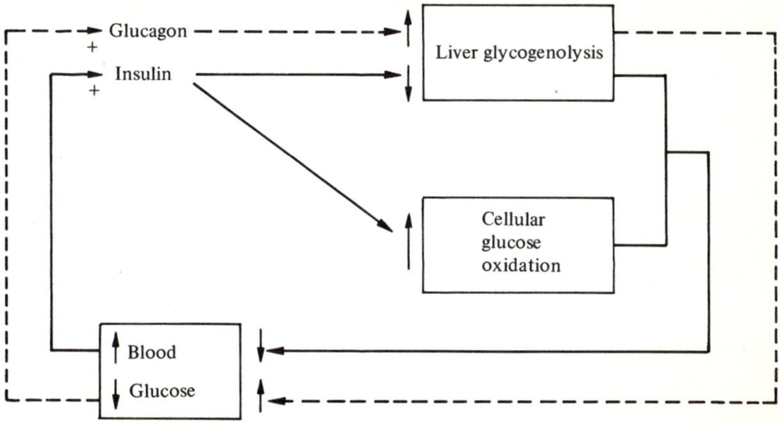

FIGURE 2-5. Diagram depicting the feedback mechanisms regulating the secretion of insulin and glucagon.

muscle and liver glycogenolysis and increased glucose oxidation lead to a lowering of blood glucose. A fall in blood glucose leads to the secretion of glucagon and an increase in blood glucose.

The availability of energy for expenditure from glycogen will depend on the factors influencing its interconversion to or from glucose. The major stores of glycogen are in the liver and in muscle. Liver glycogen is rapidly depleted to maintain blood sugar levels and liver glycogen is formed when the blood sugar level rises. In liver and muscle this balance depends on the blood level of insulin; in the liver glucagon is an additional factor. The relationship is schematically shown in a series of graphs of the transfer function between the variable on the abscissa and the parameter on the ordinate (Figure 2-6).

Thus insulin and glucagon appear to play a reciprocal role in the regulation of the blood glucose level, and the principal mechanism appears to be the increased glycogen synthesis caused by insulin and the increased glycogenolysis caused by glucagon. There are other metabolic actions attributable to insulin but this discussion will be limited to major effects in the metabolic scheme.

At the molecular level insulin appears to have a direct effect on glycogen synthesis in addition to an effect on membrane permeability to glucose. The obvious point of action is between glucose-1-phosphate and glycogen. One mechanism suggested is based on the finding that in vitro additions of insulin increase the amount of UDP-glucose-x-glucan-transglucosylase activity in the rat diaphragm. This enzyme catalyzes the formation of 1,4-alpha-glycoside chains of glycogen from UDP-glucose. Further, the enzyme exists in two forms, one requiring glucose-6-phosphate for activity. The two forms are interconvertible, and insulin is presumed to cause the formation of the glucose-6-phosphate independent form. The membrane permeability change

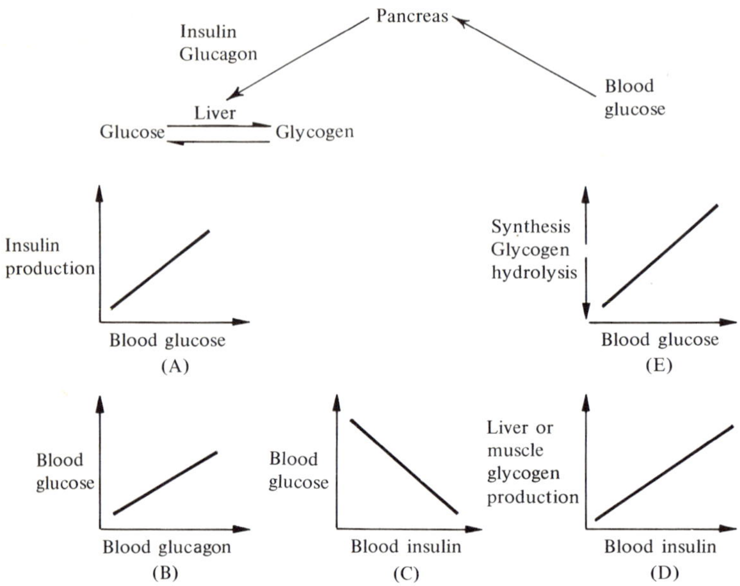

FIGURE 2-6. Series of graphs of the transfer function between the variable on the abscissa and the parameter on the ordinate. The pancreas secretes insulin and glucagon in response to the glucose concentration in the blood. A: Increased blood glucose concentration causes an increase in insulin secretion. B: An increase in glucagon in the blood gives rise to an increase in blood glucose by liver glycogenolysis. C and D: An increase in insulin in the blood causes blood glucose to decrease and causes an increase in glycogen formation in liver or muscle. E: Thus, the level of the glucose in the blood adjusts the balance between synthesis and hydrolysis of glycogen in the liver. Although these relationships are shown as linear, this is highly unlikely. The quantitive relationships are not fully known.

to glucose-6-phosphate would act in a mass action fashion to form glycogen —the increased rate of synthesis would enhance the effect.

Glucagon acts in a manner similar to epinephrine.

EPINEPHRINE

Epinephrine has a calorigenic effect in mammals (Figure 2-7). In addition, the hormone causes a marked increase in blood glucose and in blood nonesterified fatty acids. The effect on blood glucose is dependent on the increase in activity of adenyl-cyclase. The mechanism depends on a cascade sequence of conversions resulting in an increase in the amount of phosphorylase-*a* and augmentation of the conversion of glycogen to glucose-1-P (Figure 2-8).

Epinephrine (as well as norepinephrine) stimulates the liberation of free fatty acids from adipose tissue. This effect is due to the increase in activity of a specific lipase, referred to as the epinephrine-sensitive lipase. The effect

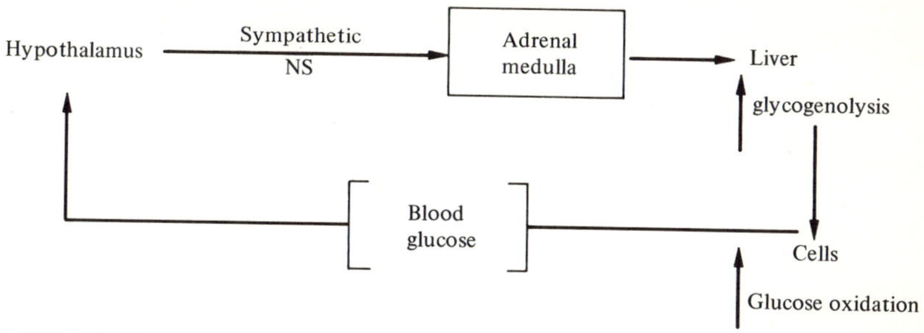

FIGURE 2-7. Feedback mechanism regulating the secretion of epinephrine.

may be mediated by cyclic 3′5′-AMP but is not dependent on phosphorylase activity. A possible tandem lipolytic effect is shown in Figure 2-9.

The calorigenic effect of epinephrine is not entirely explained, as it is not due to a forcing effect of raised blood levels of glucose or free fatty acids.

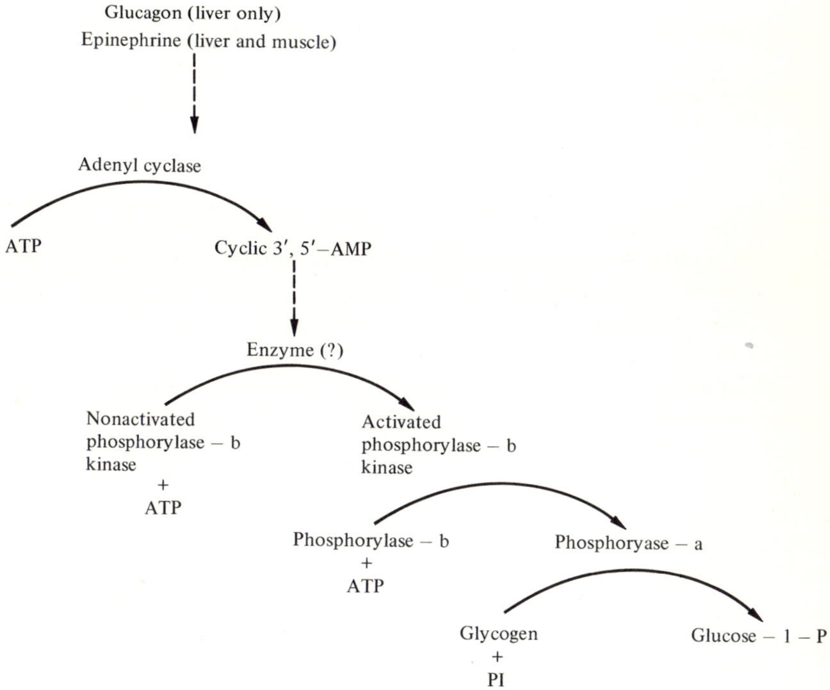

FIGURE 2-8. The mechanism of action of glucagon and epinephrine on glycogenolysis in muscle and liver. (From E. G. Krebs, R. J. DeLange, R. G. Kemp, and W. D. Riley. Activation of skeletal muscle phosphorylase. *Pharm. Rev* **18:** 163–171; 1966. © 1966, The Williams & Wilkins Co., Baltimore, Md. 21202, U. S. A.)

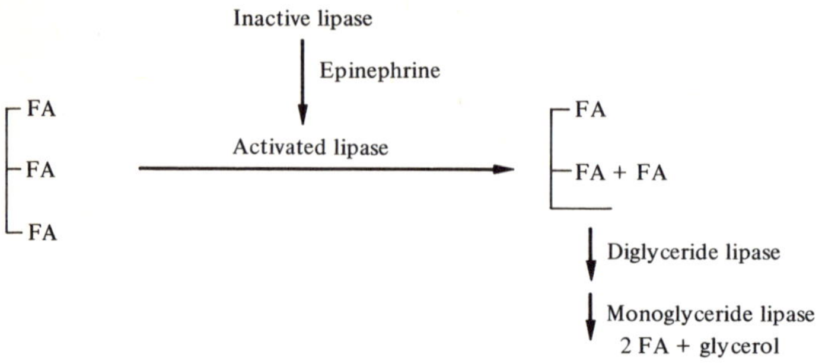

FIGURE 2-9. Possible tandem lipolytic effect of epinephrine.

CORTISOL

Cortisol is termed a glucocorticoid because of its effect on gluconeogenesis from protein (Figure 2-10). The hormone alters the levels of amino acid transaminase, so that protein synthesis is impaired by reduction in concentration of certain amino acids and gluconeogenesis is permitted to proceed at a faster rate by provision of substances such as alpha-keto-glutarate in the Krebs' cycle. Cortisol has been termed a permissive hormone because its presence is necessary for the acceleration of a reaction by other mechanisms, but it does not produce it per se.

THYROXIN AND TRIIODOTHYRONINE

The concentration of thyroid hormone in the blood is regulated by a feedback mechanism shown in Figure 2-11. The immediate stimulus for secretion of the hormone is the concentration in the blood of thyroid stimulating hormone (TSH) released from the anterior pituitary gland in response to stimuli, thyrotrophin-releasing factor (TRF), from the hypothalamus. The concentration of the circulating thyroid hormone exerts a negative feedback stimulus on the hypothalamus. Certain stressing agents, such as cold, acting

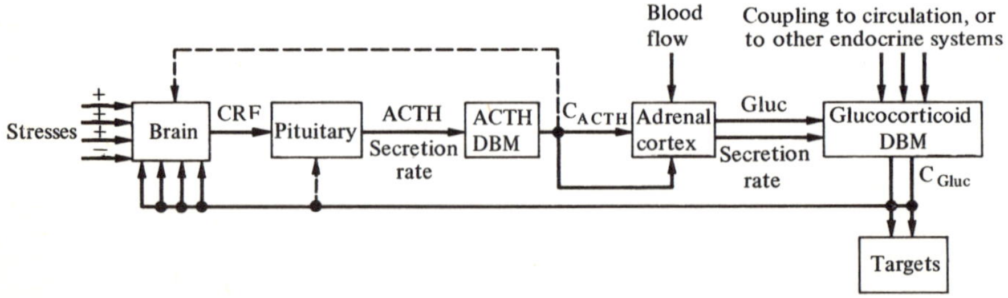

FIGURE 2-10. Feedback mechanism regulating the secretion of cortisol. DBM = distribution, binding, and metabolism elements.

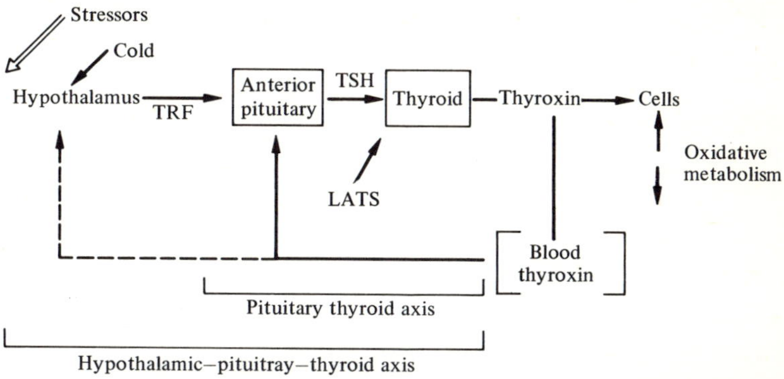

FIGURE 2-11. Feedback mechanism regulating the secretion of thyroid hormone.

through the central nervous system, can override the negative feedback system and thus raise the level of circulating thyroid hormone. In order to account for certain thyrotoxic conditions, a long-acting thyroid-stimulating hormone (LATS) has been postulated.

The power of the negative feedback system is illustrated in Figure 2-12*A*. Raising the concentration of thyroxin in the blood to about 12 μg per cent almost completely inhibits the secretion of TSH. The hypothetical effect of raising blood TSH on thyroxin level is a linear one (Figure 2-12*B*).

The hormones from the thyroid gland (thyroxin [T_4] and triiodothyronine [T_3]) have a direct effect on metabolism. Removal of the thyroid gland is followed by a gradual reduction in metabolism over a period of days to a level of 50 to 60 per cent of the resting or basal level. This metabolism can still be increased by activity but the maximum is very restricted. Injection of thyroxin leads to an increase in metabolism with a time lag, T_3 having a shorter response time than T_4. When the individual tissues are examined for their metabolic reaction to thyroxin, muscle, liver, and kidney are affected

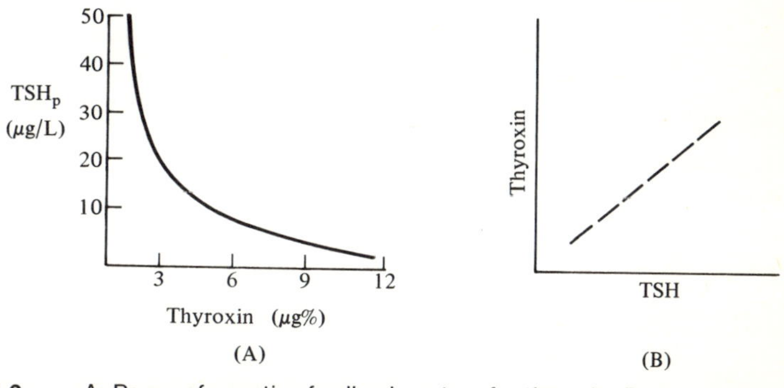

FIGURE 2-12. A: Power of negative feedback system for thyroxin; B: hypothetical linear effect on thyroxin level of raising the blood TSH.

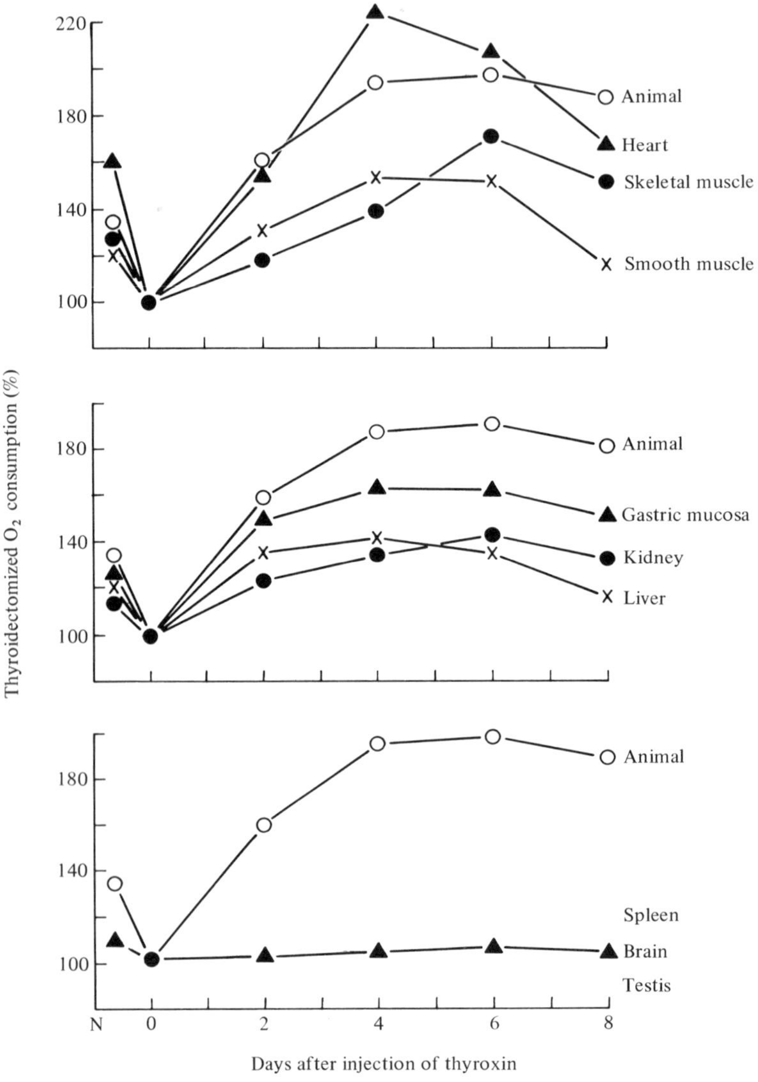

FIGURE 2-13. The effects of a single injection of thyroxin on the oxygen uptake of tissues and the whole animal of thyroidectomized rats. (From S. B. Barker and H. M. Klitgard. *Amer. J. Physiol.* **170:** 81–86, 1952.)

whereas tissues like the brain are not (3). The relative effect of thyroxin on tissues is shown in Figure 2-13. The lack of thyroxin has an inverse effect. The clinical manifestations of thyroid hormone deficiency and excess are listed in Table 2-2. Note that most of these are reflections of metabolic changes. The measurement of basal metabolism has been classically related to diseases of the thyroid, although the clinical syndrome is a complex of the manifestations shown in Table 2-2 and the current clinical tests are of the blood-protein-bound iodine (PBI) level and ^{131}I uptake and release rates. Blood-protein-bound iodine is used as an index of thyroid disease. The normal range is 4 to 8 gamma per cent. Radioiodine uptake by the thyroid is also used. An oral dose is given and the percentage of the dose in the thyroid (from an external count) at 1, 2, 6 or 24 hr may be used. The range of thyroid uptake of ^{131}I in Figure 2-14 shows the variation with disease and the wide range in normal. Note the possible error in assessment for the hyperthyroid individual with a single 24-hr count. The large variation in the normal range from 15 to 50 per cent uptake at 24 hr is unexplained. The biologic decay of thyroxin from the thyroid is also indicative of the rate of secretion from the gland. The range of biologic decay in normal man expressed as the half-time is 56 to 152 days. These wide ranges in function have not been correlated with metabolic measurements.

At the cellular level, thyroxin has been said to be an "uncoupler" in phosphorylation of ADP and has been characterized as a "calorigenic shunt." This hypothesis is based on the observation that in simple metabolic systems the rate of oxidation of substrate, glucose, appears to be geared to the

TABLE 2-2
Manifestations of Thyroid Hormone Deficiency and Excess*

	Excess	Deficiency
Metabolism	Increased	Decreased
Basal Metabolic Rate (BMR)	(to +50 per cent)	(to −30 per cent)
Muscle	Weakness, tremors	Weakness, hypotonia
Sweating	Increased	Decreased
Body weight (adult)	Decreased	Increased or no change
Growth (in young)	Rapid	Cessation (dwarfism)
Tissue changes	Exopthalmos	Myxedema
Temperature sensitivity	Sensitivity to heat	Sensitivity to cold
Serum cholesterol	Normal or low	Elevated
Serum PBI	15 μg per cent	<2 μg per cent
Heart rate	Increased	Decreased
Pulse pressure	Increased	Decreased
Cardiac output	Increased	Decreased
Gastrointestinal response	Diarrhea	Constipation (flatulence)
CNS response	Nervousness, tremor	Irritability, deafness
Reflex	Hyperactive	Slow
Cerebration	Increased	Decreased
Drug tolerance	Increased	Decreased
Skin	Increased sweating, hot, flushing	Dry, scaly, cold, thickened
Hair	Fine, soft, straight, temporary loss	Dry, brittle, sparse

* Prepared by Dr. L. K. Bustad.

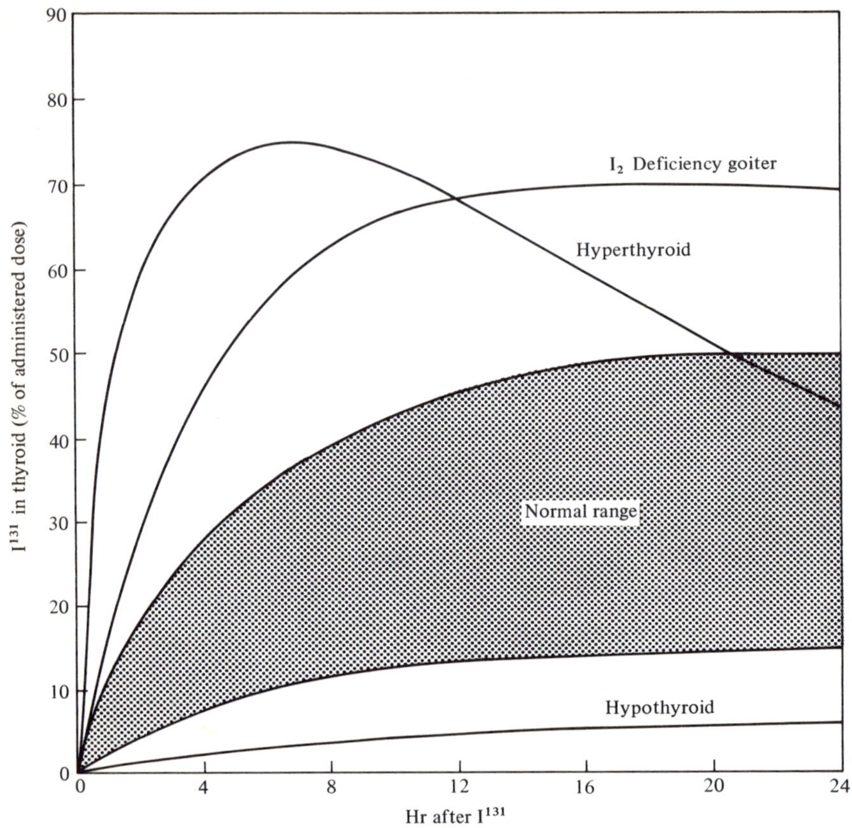

FIGURE 2-14. Normal range of thyroid uptake in man as compared with typical hypo- and hyperthyroid uptake curves obtained from many sources.

amount of phosphate acceptor, ADP, present. In the presence of thyroxin the rate is increased and oxidation of substrate appears to be uncoupled from the ADP–ATP reaction. The hormone increases the effective concentration of a number of enzymes and changes the nature of the mitochondrial membrane. Another current hypothesis concerning the action of thyroxin postulates a general influence on protein synthesis leading to a general increase in enzyme concentrations.

In the hypothyroid rat energy transfer in the liver mitochondria is suppressed through excessive respiratory control, and a very small dose of 1-thyroxin restores in 3 hr normal respiratory control of substrates initially oxidized by pyridine-nucleotide-dependent enzymes. In the normal rat a larger dose of hormone increases mitochondrial respiration in 6 hr. These functional changes appear to be due to a direct action of the administered hormone on the liver mitochondria, but it is not yet certain which of the components of the mitochondrial apparatus change their function because the hormone is present or which change their function because of earlier mitochondrial, structural, functional, or compositional changes induced by the hormones. The following series of events appear reasonable as a working hypothesis in hypothyroidism and following thyroid administration. Continued depression of mitochondrial energy transfer in the untreated hypothyroid rat results in a depression of synthetic processes. The known decrease in protein synthesis produces mitochondrial respiratory assemblies with too few cytochrome enzymes. This deficiency in enzyme content seems to be a specific one, perhaps brought about by the low rate of utilization of mitochondrial substrate oxygen in hypothyroid mitochondria—in much the same way that depriving bacteria of substrates can suppress the synthesis of the enzymes that metabolize these substrates. There is evidence that oxygen alters the constitution of the electron transport systems of mitochondria. Excessive concentrations of oxidized pyridine nuclides are found in the tissues of animals exposed to high oxygen pressures. The administration of thyroid hormone to hypothyroid rats causes a series of events. Energy transfer is increased in mitochondria. The increased supply of high-energy compounds releases the depressed ribosomal interaction

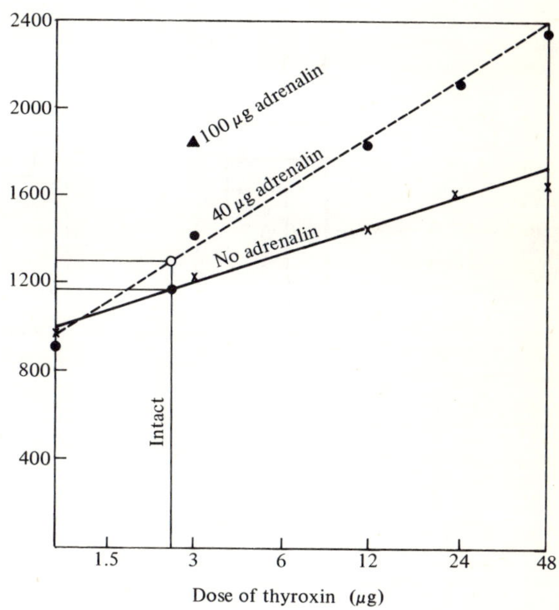

FIGURE 2-15. Regression of oxygen consumption of thyroidectomized rats treated with various doses of thyroxin plotted against the dose of thyroxin on a logarithmic scale. Without adrenalin, the slope was 0.31. One hour after 40 μg of adrenalin per 100 gm body weight, the slope was 0.62. A dose of 100 μg was fatal to all animals receiving more than 3 μg of thyroxin. (From H. E. Swanson. *Endocrinol.* **60:** 205, 1957.)

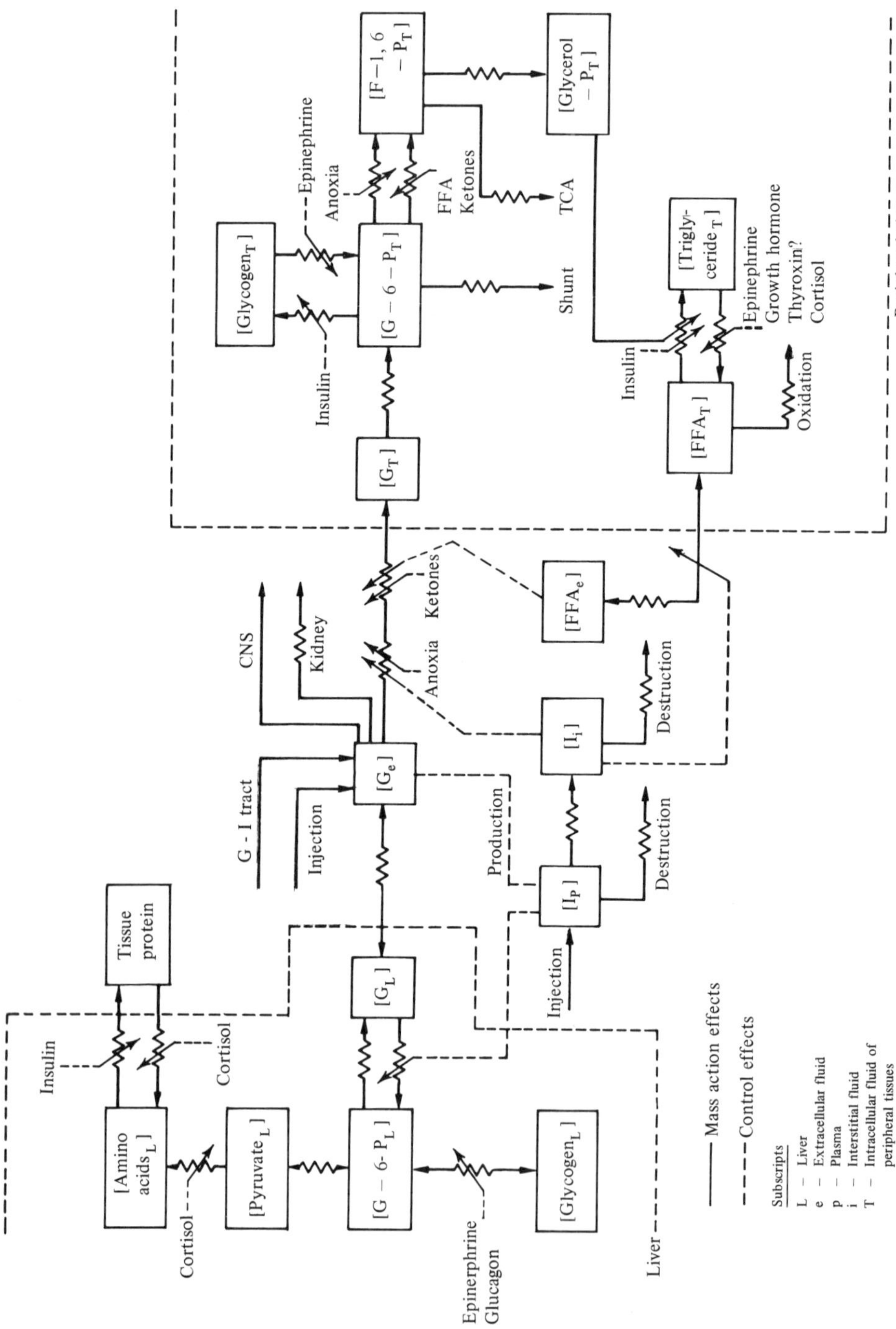

FIGURE 2-16. Hormonal effects on glucose regulation. (From D. M. Shanes. A theoretical study of the blood glucose regulatory system. M. D. Thesis, Yale University, 1965.)

with RNA–amino acid complexes. The increased rate of substrate utilization and the functional alterations of structure in the metabolic state of the mitochondria control its protein synthesis and stimulate a specific synthesis of mitochondrial components.

THYROXIN SYNERGISM WITH EPINEPHRINE

Thyroxin action on metabolism in the animal is synergistic with that of epinephrine (14). The data obtained by studies on white rats (Figure 2-15) illustrate this relationship. This relationship has not been explored for norepinephrine nor is the relationship at the tissue level known. It is tempting to suggest that the synergistic action results from the specific action of epinephrine at a rate-limited process on the metabolic machinery where other enzyme levels may be thyroxin dependent.

SUMMARY

The hormonal effects on the metabolism of glucose (13) are summarized in Figure 2-16.

References

1. Adolph, E. F. Urges to eat and drink in rats. *Amer. J. Physiol.* **151:** 110–125, 1947.
2. Aschoff, J., and R. Wever. Kern und schale in warmhaushalt des menschen. *Naturwissenschaften* **45:** 447–485, 1958.
3. Barker, S. B., and H. M. Klitgaard. Metabolism of tissues excised from thyroxine-injected rats. *Amer. J. Physiol.* **170:** 81–86, 1952.
4. Boothby, W. M., J. Berkson, and W. A. Plummer. The variability of basal metabolism: Some observations concerning its application in conditions of health and disease. *Ann. Intern. Med.* **11:** 1014–1023, 1937.
5. Chance, B. In: *Regulation of Cell Metabolism.* Edited by G. E. W. Wolstenholme and C. M. O'Connor. Boston: Little, Brown, 1959, p. 355.
6. Engelberg, J. By way of introduction—The cell. In: *The Physiological Basis of Medical Practice,* 8th ed. Edited by C. H. Best and N. B. Taylor. Baltimore: Williams and Wilkins, 1966, pp. 1–21.
7. Goldberg, M., and E. S. Gordon. Energy metabolism in human obesity, plasma free fatty acid, glucose and glycerol response to epinephrine. *JAMA* **189:** 616–623, 1964.
8. Gordon, E. S., M. Goldberg, and Grace Chosey. A new concept in the treatment of obesity. *JAMA* **186:** 50–60, 1963.
9. Mayer, J. Gross efficiency of growth of the rat as a simple mathematical function of time. *Yale J. Biol. Med.* **21:** 415–419, 1948–49.
10. Pace, N., and E. N. Rathbun. Studies on body composition. III. The body water

and chemically combined nitrogen content in relation to fat content. *J. Biol. Chem.* **158:** 685–691, 1945.

11. Rathbun, E. N., and N. Pace. Studies on body composition. I. The determination of total body fat by means of body specific-gravity. *J. Biol. Chem.* **158:** 667–676, 1945.
12. Reichlin, S., and R. D. Utiger. Regulation of the pituitary-thyroid axis in man: Relationship of TSH concentration to concentration of free and total thyroxine in plasma. *J. Clin. Endocr.* **27:** 251–255, 1967.
13. Shanes, D. M. A theoretical study of the blood glucose regulatory system. M. D. Thesis, Yale University, 1965.
14. Swanson, H. E. The effect of temperature on the potentiation of adrenaline by thyroxine in the albino rat. *Endocrinology* **60:** 205–213, 1957.
15. Woodbury, D. M. Physiology of body fluids. In: *Physiology and Biophysics.* Edited by T. C. Ruch and H. D. Patton. Philadelphia: Saunders, 1965, p. 888.
16. Yates, F. E., and R. D. Brennan. Study of the mammalian adrenal glucocorticoid system by computer simulation. International Business Machines Technical Report #320–3328, 1967.

Additional Reading

Gemmell, C. L. Energy metabolism. In: *Medical Physiology.* 12th ed. Edited by V. B. Mountcastle. St. Louis: Mosby, 1968, chap. 30.

Jöbsis, F. F. Basic processes in cellular respiration. In: *Handbook of Physiology. Sect. 3. Respiration. Vol. I.* Edited by W. O. Fenn and H. Rahn. Washington: American Physiological Society, 1964, pp. 63–124.

Litwack, G., and D. Kritchevsky. *Actions of Hormones on Molecular Processes.* New York: Wiley, 1964.

Tepperman, J. *Metabolic and Endocrine Physiology.* Chicago: Yearbook Medical Publishers, 1962.

Wostenholme, G. E. W., and C. M. O'Connor. *Regulation of Cell Metabolism.* Boston: Little, Brown, 1958.

CHAPTER

3

Expenditure of Energy

Energy expenditures are required for active transport of material across cell membranes, secretion, synthesis, and mechanical work of smooth and striated muscle.

Active Transport

An example of active transport of inorganic ions is the maintenance of an ion gradient between the interior and exterior of cells. The concentration of Na^+ inside the cells is lower than that in the extracellular fluid. Since the cell membrane is permeable to Na^+, and since the interior of the cell is electrically negative with respect to the exterior, it follows that Na^+ is extruded against an electrical potential gradient. A "sodium pump," using energy supplied by the hydrolysis of ATP, has been postulated to be situated in the cell membrane. Ussing (12) has calculated the energy required to maintain the electric potential gradient in muscle to be 50 cal/kg of muscle water per hr. This is about 10–20 per cent of the resting metabolism of muscle.

The absorption of sugar by epithelial cells of the intestine is apparently dependent only on a continuously maintained outward Na^+ gradient (2). The energy required for sugar accumulation in mucosal cells is thus linked with that which is necessary to maintain the proper functioning of the "sodium pump." A current hypothesis postulates the activation by Na^+ of a mobile sugar carrier which is situated in the brush border on the mucosal membrane.

The presence of K^+ decreases the affinity of the carrier for sugar. Thus the carrier picks up sugar at the luminal border and dumps sugar at the inner border where Na^+ concentration is lower and K^+ concentration higher.

Secretion

The secretion of HCl by the parietal cells of the stomach is a good example of energy expenditure for secretory processes. The cells raise the H^+ concentration from 0.00005 mN in plasma to 150 mN in gastric juice. The secretion of H^+ is associated with the secretion of Cl^- which also moves against a concentration gradient. A current hypothesis postulates electrogenic chloride and proton pumps coupled by some unknown means. The energy required for the secretion of HCl is estimated to be at least 1.5 kcal/l of juice (1).

Synthesis

All biosynthetic reactions are endergonic processes, that is, they consume energy, and cannot take place unless energy is supplied to the system. The energy is derived from ATP either directly or indirectly. During protein synthesis the cell uses ATP to produce a reactive intermediate, amino acyl sRNA, which transfers the amino acid to the messenger RNA and ribosomal RNA for polymerization.

An example of synthesis that is of interest in the context of this monograph is the formation of fatty acids from acetyl-Co-A. The formation of one mole of palmitic acid uses about seven moles of ATP. Conversion back to acetyl-Co-A requires a further mole of ATP. Thus the fatty acid synthesis-oxidation cycle could act as an ATP utilizing heat generator. Since catabolism of glucose via fatty acid synthesis and oxidation leads to less ATP formation and more heat liberation than the direct route, this lipogenic pathway has been called a "calorigenic shunt." Activation of this pathway could account for non-shivering thermogenesis (9).

Contraction of Muscle

The action of striated muscle has been modeled from the structural mechanism proposed by Huxley (Figure 3-1). Two proteins, actin and myosin, and ATP are the principal participants in the contractual process. The arrangement of the filaments are such that cross bridges link thick filaments and thin filaments, allowing them to be moved by a ratchetlike action. Practically all the ATP in muscle is bound to myosin. The transfer of energy from ATP to contraction—motion or tension—may be related to cross bridges bending back and forth or the coupling and uncoupling with active sites.

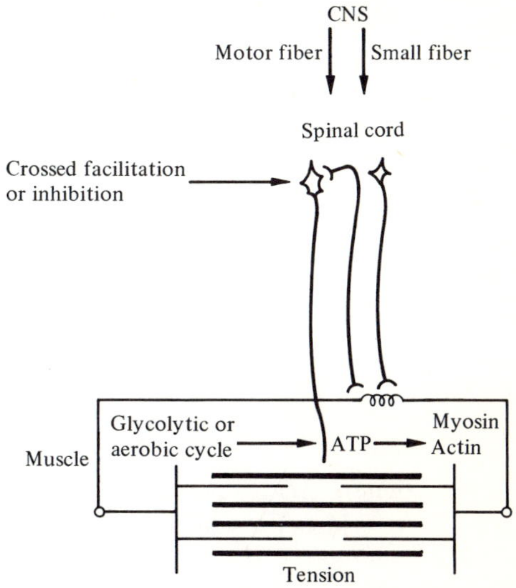

FIGURE 3-1. The central nervous system (CNS) controls muscle contraction via neural pathways. Release of transmitter substances at the nerve endings leads to the transfer of energy from ATP to contraction.

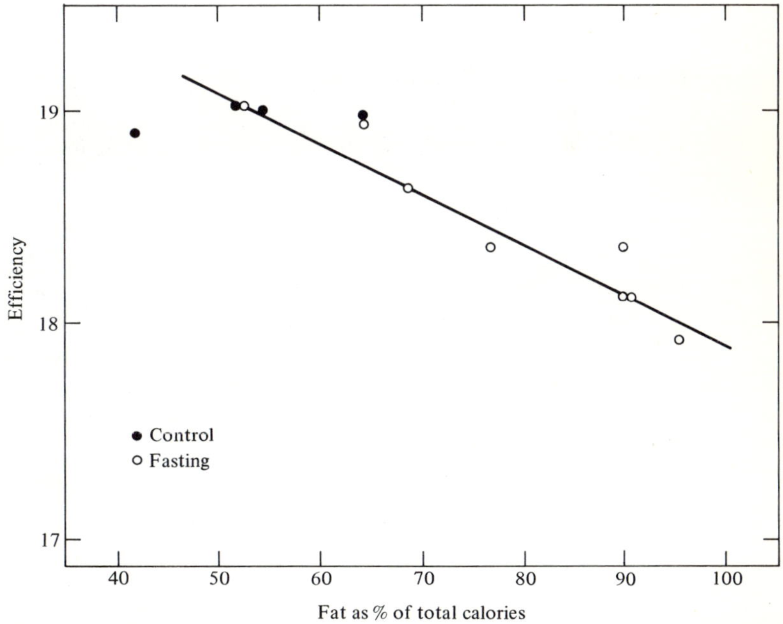

FIGURE 3-2. The relationship of mechanical efficiency of work in grade walking to the per cent of total calories burned as fat during acute starvation with work. (From A. Henschel, H. L. Taylor, and A. Keys. *J. Appl. Physiol.* **6:** 624–633, 1954.)

Work

The strength of isometric (tension) and isotonic (shortening) contractions of muscle is related to the length for isometric contraction and to load in isotonic contraction. Only in isotonic contraction is work done. Maximum work is at one half maximum load. In isotonic contraction

$$\Delta L = \Delta L_{max}\left(1 - \frac{F}{F_{max}}\right)$$

$$\text{Work} = F\Delta L_{max}\left(1 - \frac{F}{F_{max}}\right)$$

where ΔL is contraction and F is load.

Muscles work at a large mechanical disadvantage. The mechanical advantage of the human biceps is 0.1 and the range for muscles is 0.05 to 0.4. Thus a 70-kg man who can chin himself with one hand is lifting with a force of 700 kg (about $\frac{3}{4}$ ton).

The efficiency of work may be as high as 20 per cent. The relationship is dependent on the caloric sources, decreasing as the per cent of calories from lipid increase (Figure 3-2).

Chemical Energy for Contraction

A. V. Hill (3) formulated the energetics for muscle in an empirical formula.

$$\text{Total chemical energy expended} = A + ax + px$$

where A is an amount of heat independent of both the amount of work done and the amount of shortening (called the heat activation);
ax is the heat of shortening, linearly related to the shortening X;
px represents the mechanical work.
The energy for contraction may be considered to involve:

1. $ATP \longrightarrow ADP + P_i$.
2. $CP + ADP \longrightarrow ATP + C$.
3. Aerobic or anaerobic rephosphorylation of creatinine.

The simplest link to oxygen consumption would be the stimulus of the increase in phosphate acceptor, ADP or C.

Mechanochemical Contraction Models

Podolsky (7) has succinctly summarized the picture of muscular contraction. In active muscle chemical energy is continuously being converted into mechanical energy and heat. While shortening, muscle generates a force that just matches the load. The relation between force and velocity characterizes the operation of the contractile mechanism. If viscous forces are not important (the heat measurements of Fenn and of Hill indicate that this is the case), the force velocity relation is

$$(P + a)V = b\,(P_0 - P)$$

where $P =$ contractile force Q,
$P_o =$ maximum force Q, and
$V =$ velocity;
a and b are constants for the heat of shortening and the rate of the driving chemical reaction, respectively. This relation implies that the sequence of events underlying contraction form a mechanochemical cycle. In other words, the energy is converted directly to contractile energy. A steady motion is generated by the interaction of a time-dependent, force-generating process (coupled to the driving chemical reaction) and a displacement-dependent, force-dissipating process.

The mechanochemical cycle, as well as other physiologic properties of living muscles, can be simulated by models composed of a double array of myosin filaments. The sliding version has been analyzed in detail by A. F. Huxley (5). A series of short elastic links connect specific sites on the thin filament and complementary sites on the thick filament. As the thick and thin filaments slide past each other, attachment sites are connected to and disconnected from pulling sites on the other. A folding mechanism has been proposed by R. J. Podolsky (7). In this model the ends of the thin filament are fixed on activation and the contractile force is generated by the tendency of this thin filament to shorten by folding. The main feature of the model is that the thin filament has a series of sites to which substrate (supplied by the driving chemical reaction) can bind, and the force is proportional to the number of "full" sites.

Relation Between Lactate and Pyruvate

The time course of the energy requirement in exercise and the rate at which the requirement is met depend on the rate at which the work is performed. Figure 3-3, from work in A. V. Hill's laboratory (1), illustrates

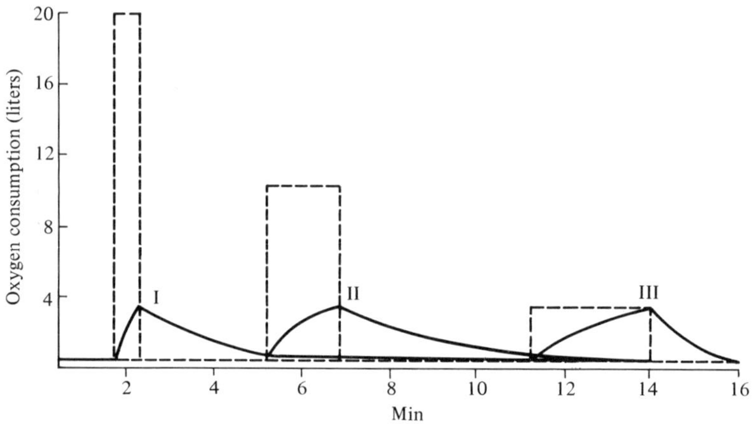

FIGURE 3-3. The difference between the oxygen requirement and the oxygen uptake. Areas under broken lines are oxygen requirement. Areas under solid lines are oxygen uptake. I: severe exercise; II: moderate exercise; III: light exercise. (From K. Furusawa, A. V. Hill, C. N. H. Long, and H. Lupton. *Proc. Roy. Soc.* [*Biol.*], **97:** 167, 1925.)

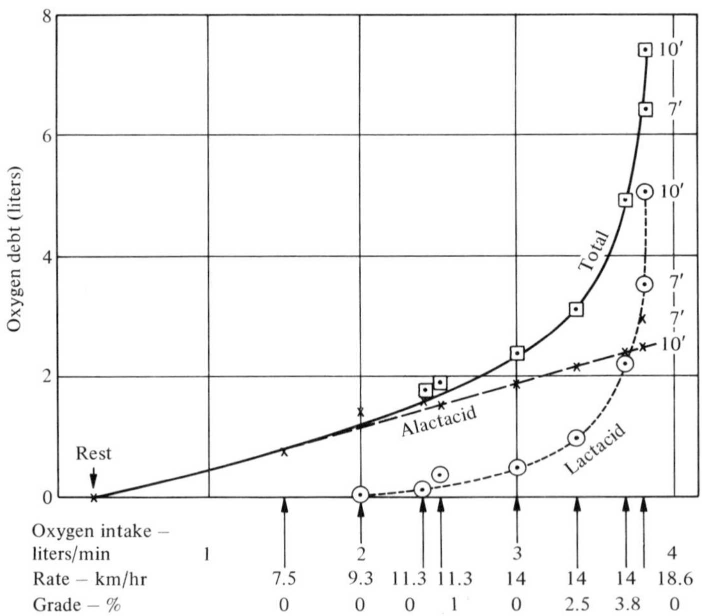

FIGURE 3-4. Relationship between lactic acid concentration in the blood and oxygen debt. Duration of exercise was 10 min. in each case. (From R. Margaria, editor. *Exercise at Altitude.* New York: Exerpta Medical Foundation, 1967.)

this for three grades of work. The lag in oxygen consumption to meet a requirement is called the oxygen deficit; the delay in repayment is the oxygen debt. A portion of the oxygen debt is the lag in reconstituting the high-energy phosphate pool. In addition to this, there is an anaerobic debt, which is reflected in the blood level of lactic acid. This debt is dependent on the amount of exercise or the energy requirement (Figure 3-4). When the lactate debt in terms of oxygen debt is subtracted from the total, the major part of the remainder is repaid at a rate 30 times that of the lactate. A minor component of the debt is repaid over a very long time period.

Lactate is formed from pyruvate following the nominal equation:

$$\text{Pyruvate} + \text{DPNH}_2 \rightleftharpoons \text{lactate} + \text{DPN}$$

In the normal course of aerobic metabolism, one would expect the lactate level to reflect the rate at which pyruvate is being mobilized. In anaerobic situations the lactate is generated to replenish the hydrogen acceptor DPN. Thus,

$$\text{Lactate} = \text{pyruvate} \times \text{K}\frac{\text{DPNH}_2}{\text{DPN}}$$

Huckabee (4) has derived an equation for calculating the fraction of change in lactate concentration that could be expected to be due to oxygen lack

$$XL = (L_n - L_o) - (P_n - P_o)\frac{L_o}{P_o}$$

$$= L_n - \frac{P_n L_o}{P_o}$$

where $L_n P_n$ = experimental lactate and pyruvate,
$L_o P_o$ = control lactate and pyruvate, and
XL = excess lactate.

"Negative" Work

Oxygen consumption increases as work output increases. In conventional physical terms work is done when a weight is lifted vertically over a distance. When the weight is lowered again, the work is zero. However, a muscle also exerts force over a distance X to lower a weight. Physiologists term work done to raise a weight positive work and that done to lower it negative work. Asmussen studied the work of bicycling up a grade and down a grade. His

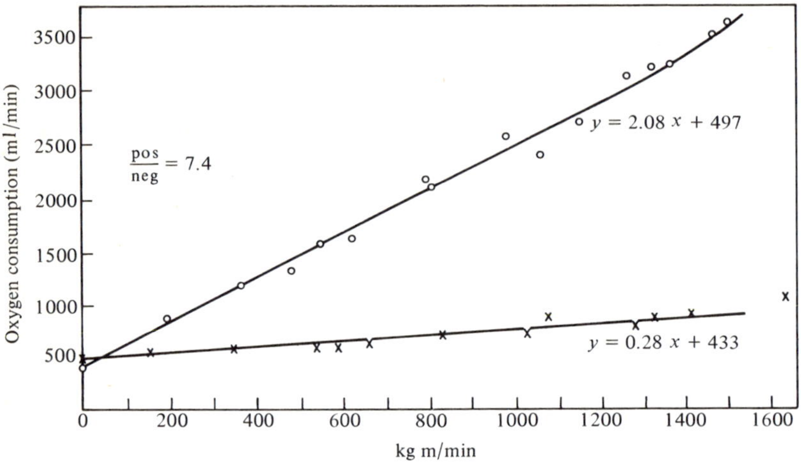

FIGURE 3-5. The oxygen consumption in milliliters per minute of a man bicycling uphill and downhill on a motor-driven treadmill plotted against the rate of work. (From E. Asmussen. Experiments on positive and negative work. In: Ergonomics Society Symposia on Fatigue. Edited by W. F. Floyd and A. T. Wilford. London: Lewis, 1953, pp. 77–83.)

data provide the relationship of oxygen consumption to rate of work and establish a ratio of positive to negative work of 7.4 (Figure 3-5). The range from many studies is 3 to 9.

References

1. Crane, R. K. Absorption of sugars. In: *Handbook of Physiology, Sec. 6, Alimentary Canal, Vol. III.* Edited by C. F. Code. Washington, D. C.: American Physiological Society, 1968, Chap. 69.
2. Davenport, H. W. Gastric Secretion. In: *Physiology of the Digestive Tract.* Chicago: Year Book Medical Publishers, 1966, Chap. 8.
3. Furusawa, K., A. V. Hill, C. N. H. Long, and H. Lupton. Muscular exercise and oxygen requirement. *Proc. Roy. Soc.* (*Biol.*) **97:** 167–176, 1925.
4. Henschel, A., H. L. Taylor, and A. Keys. Performance capacity in acute starvation with hard work. *J. Appl. Physiol.* **6:** 624–633, 1954.
5. Hill, A. V. The heat of shortening and the dynamic constants of muscle. *Proc. Roy. Soc.* (*Biol.*) **126:** 136–195, 1938.
6. Huckabee, W. A. Relationships of pyruvate and lactate during anerobic metabolism and effects of infusions of pyruvate or glucose and of hyperventilation. *J. Clin. Invest.* **37:** 244–254, 1958.
7. Huxley, A. F. Muscle structure and theories of contraction. *Progr. Biophys.* **7:** 255–318, 1957.
8. Margaria, R., H. T. Edwards, and D. B. Dill. The possible mechanisms of contracting and paying the oxygen debt and the role of lactic acid in muscular contraction. *Amer. J. Physiol.* **106:** 689–715, 1933.

9. Masoro, E. J. Role of lipogenesis in nonshivering thermogenesis. *Fed. Proc.* **22:** 868–873, 1963.
10. Podolsky, R. J. Mechanochemical basis of muscular contraction. *Fed. Proc.* **21:** 964–974, 1962.
11. Taylor, H. L. In: *Science and Medicine of Exercise and Sports,* edited by W. R. Johnson, New York: Harper & Row, 1960, Chap. 8.
12. Ussing, H. H. Active transport of inorganic ions. *Symp. Soc. Exptl. Biol.,* **8:** 407–422, 1954.

Additional Reading

Ackerman, E. In: *Biophysical Science.* Englewood Cliffs, N.J.: Prentice-Hall, 1962, Chap. 8.

Davson, H. A. *Textbook of General Physiology.* Boston: Little, Brown, 1964, Sec. 5 and Chap. 8.

Edholm, O. G. *The Biology of Work.* London: Wiedenfield and Nicolson, 1967.

Johnson, W. R., ed. *Science and Medicine of Exercise and Sports.* New York: Harper & Row, 1960.

CHAPTER

4

Thermophysiology

Over-all Equation

Thermophysiology involves the physiologic measurements that reveal energy exchange between the organism and its environment. Consider a body or body element as shown in Figure 4-1. This body will have a heat production M, which will necessarily be distributed within the body and transferred to the surface for dissipation to the environment. A portion of this body may change in temperature, thus contributing or absorbing heat as a part of the over-all exchange H. Storage S has the same algebraic sign as M when the body is cooling. The over-all equation for heat exchange is

$$H = M \pm S = E \pm K \pm C \pm R + W$$

where E = evaporative heat loss,
K = conductive heat exchange (negative for heat gain by body),
C = convective heat exchange (negative for heat gain by body),
R = radiant heat exchange (negative for heat gain by body), and
W = external work accomplished.

When the body is at rest, $W = 0$, and in thermal equilibrium, $S = 0$, heat loss from the skin surface is

$$H = M = h_{kcrs}(T_s - T_a) + E$$

where h_{kcrs} = heat transfer coefficient for K, C, and R,

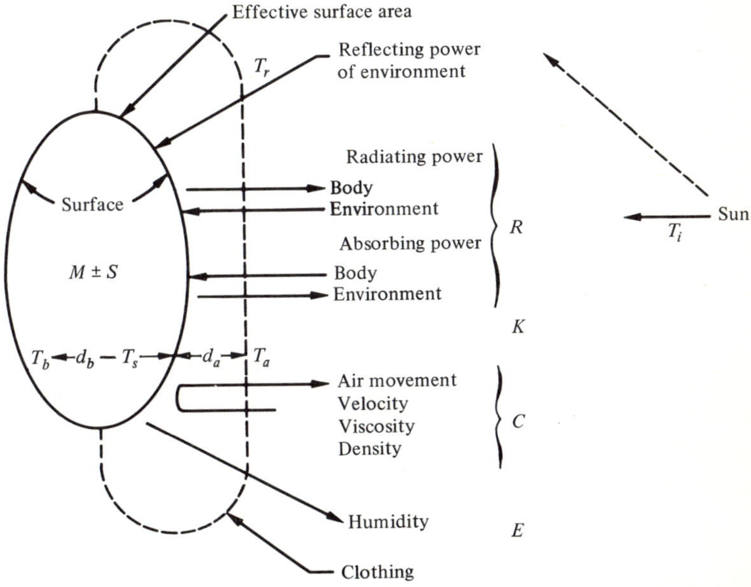

$M \pm S = \pm R \pm K \pm C + E$

FIGURE 4-1. Model of a body element exchanging heat with the environment.

E = g of water evaporated X heat of evaporation at T_s,
T_s = skin temperature, and
T_a = air temperature.

Heat loss from the interior of the body is

$$H = M = h_{kcb}(T_b - T_s)$$

where h_{kcb} = heat transfer coefficient for conductive and convective loss from the interior to skin,
T_b = body temperature.

This sequence of equations formulates the discussion of factors of heat production and heat transfer from the body. The discussion of this general diagram will be separated into the skin-to-environment system and the body tissue-to-skin system.

Skin-to-Environment System

CONDUCTION

Two basic laws are involved in consideration of nonevaporative heat loss from the body. Fourier's law describes heat flow:

$$\frac{M}{t} = \frac{A}{r_1 d}(T_b - T_s)$$

and

$$\frac{M}{t} = \frac{A}{r_2 d}(T_s - T_a)$$

In these equations, d is the distance between T_b and T_s (or T_s and T_a); r_1 is the thermal resistance, which is composed of the reciprocal of tissue conductivity and the convective loss provided by circulation of blood to the skin; and r_2 is a composite of a pelt or clothing insulation and the insulation of air. Because A, the area, is proportional to $W^{2/3}$ (W = body weight in kg), the heat loss is directly related to body weight.

Newton's law of cooling is applicable when the body is cooling. The equation is

$$\frac{S}{t} = \frac{A}{r_2 d}(T_s - T_a)$$

because S is tissue cooling, $S = cW \cdot \Delta T$. Thus,

$$\frac{dT}{dt} = \frac{A}{cWr_2 d}(T_s - T_a)$$

As stated earlier, $A \propto W^{2/3}$ and so the cooling rate will be inversely proportional to $W^{1/3}$.

Change in temperature is associated with a change in entropy. The change in entropy from central to skin temperature, considering the irreversible heat flow, would be

$$S = \frac{Q}{T_b} - \frac{Q}{T_s}$$

and

$$\frac{dS}{dt} = \frac{dQ}{dt}\left(\frac{T_b - T_s}{T_b T_s}\right)$$

The establishment of a mean body tissue temperature would permit an estimation of entropy change caused by change in tissue temperature.

As the body warms or cools, peripheral tissue temperature changes to varying depths. It is apparent that there are temperature changes not only in the skin but in the underlying tissues as well. Because approximately 50 per cent of the body mass is within 2.5 cm of the surface, a significant mass is involved in heat exchange. Weighting factors for determining mean body temperature have been determined empirically by Burton as $0.4T_s$ and $0.6T_b$ (6), and by Hardy as $0.2T_s$ and $0.8T_b$ (12, 13).

Conduction may occur to air or through added insulation such as the animal's pelt or human clothing. In either case the relationship is

$$\frac{K}{t} = \frac{A}{r_2 d}(T_s - T_a)$$

which represents the heat loss per unit time. When K, A, d, and $(T_s - T_a)$ have the units cal/sec, cm^2, cm, and °C, respectively, r_2 (the thermal resistance or insulation) will have the unit °C · cm^2 · sec/cm · cal. The reciprocal of r_2 is conductivity, with the unit cal · cm/cm^2 · sec · °C.

Some representative conductivities are

	cal · cm/cm^2 · sec · °C
Air	0.00057
Helium	0.00035
Animal fat	0.00049
Animal muscle	0.00067 to 0.00092
Water	0.001429

Calculations of thermal resistance or conductivities in mammalian systems usually ignore the dimension of length d; thus changing values may actually represent changes in this dimension as well as conductivity.

CONVECTION

Convective heat loss C is related to air or fluid movement. The equation that best fits the experimental facts and theoretical considerations is

$$C = \frac{h_c}{d}\left[1 + a\left(\frac{dV\rho}{\mu}\right)^{1/2} + b\left(\frac{dV\rho}{\mu}\right)\right](T_s - T_a)$$

where d = characteristic dimension of the object (for example, diameter of a sphere or cylinder),
V = velocity of the air or fluid,
ρ = density of the air or fluid,
μ = viscosity of the air or fluid, and a and b are constants depending on the units used.

Neglecting free convection and $b(dV\rho/\mu)$, which are small, and rearranging, we obtain

$$\frac{Cd}{h_c} = \left[a\left(\frac{dV\rho}{\mu}\right)^{1/2}\right](T_s - T_a)$$

Because Cd/h_c is Nusselt's number (Nu) and $dV\rho/\mu$ is Reynold's number (Re) for forced convection:

$$\mathrm{Nu} = a\mathrm{Re}^{1/2}(T_s - T_a)$$

Buettner plotted Nu against Re for spheres of several diameters and found $a = 0.70$ (for V greater than 0.2 m/sec). For men lying on their backs on the floor, he arrived at

$$C = 0.021V^{1/2}(T_s - T_a)$$

RADIATION

Radiant heat loss or gain is

$$R = \sigma E_s E_e (T_s^4 - T_e^4)A$$

where σ = Stefan-Boltzmann constant (5.67×10^{-8} W/m$^2 \cdot$ °K^4),
$E_s E_e$ = emissibility of objects,
T = absolute temperature, and
A = effective radiating area.

Kirchhoff's radiation law states that the ratio between the absorptivity and emissive power is the same for each kind of ray for all bodies at the same temperature and is equal to the emissive power of a black body at that same temperature. At skin temperature (about 300° K) E_s is independent of visible color and is more than 0.9. Below 1.0μ wavelength the reflecting power is color dependent.

If $(T_s - T_e)$ is less than 20°K, the radiation exchange equation can be simplified to:

$$R = h_r(T_s - T_e)$$

where h_r is the radiant-heat-transfer coefficient which equals $4\sigma T^3 A$ in which T is the mean of T_s and T_e in °K.

EVAPORATION

Evaporative loss is determined by the heat of vaporization of water, and this is temperature dependent. Some representative values are the following:

°C	cal/g
0	595.9
20	584.9
30	579.5
40	574.0

The amount of evaporation is related to the water content of the air (humidity) and the atmospheric pressure. Thus, evaporative loss at 1 atmosphere pressure will be

$$E = h_e(p_{ws} - \phi_a p_{wa})A$$

where h_e = heat-transfer coefficient by evaporation at T_s and is a function of air movement, viscosity, density, and thermal conductivity;
p_{ws} and p_{wa} = vapor pressure at T_s and T_a;
A = effective evaporating surface area; and
ϕ_a = relative humidity.

Empirical relationships that have been established vary. The Nelson equation is

$$E = 1.4V^{0.37}(p_{ws} - \phi_a p_{wa})$$

whereas the Clifford equation is

$$E = 0.45V^{0.63}(p_{ws} - \phi_a p_{wa})$$

where V is the velocity of air movement.

EVAPORATIVE LOSS IN RESPIRATION

Air exchanged in pulmonary respiration is rapidly warmed to body temperature, or cooled in a hot environment, and saturated with water vapor at body temperature during inspiration. During expiration the reverse process occurs. The actual heat exchange can be determined by measurements of volume, temperature change, and water vapor change. In the general system described, the insensible loss which includes respiration has been estimated to be 24 per cent of the total heat loss. Both the obligatory insensible water loss from the skin (insensible perspiration) and respiratory water loss depend on temperature and relative humidity in the environment. Respiratory water loss also depends on the respiratory minute volume. An estimate of these losses is shown in Figure 4-2.

GENERAL CONSIDERATIONS AND APPROXIMATIONS

For general considerations of heat loss from the body a number of approximations are used. A general equation for nonevaporative heat loss is

$$R + C = h(T_s - T_a)$$

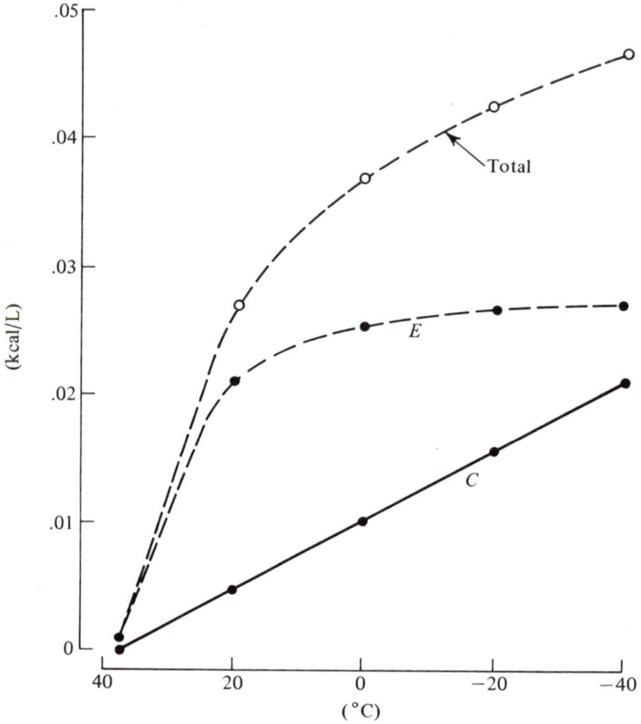

FIGURE 4-2. Estimate of heat lost to air respired at different temperatures at 50 per cent relative humidity. Expired air temperature is assumed to be at body temperature as there are insufficient data to make the necessary corrections.

In this equation, h is a combined-heat-transfer coefficient and T_a is the result of the combination of a number of factors, such as air movement and radiation surface, with air temperature. There have been numerous proposals for methods of combining these factors; a discussion of some of these follows.

TEMPERATURE RESULTANTE AND OPERATIVE TEMPERATURE. Gagge derived an operative temperature T_o from the following sequence of equations:

$$R + C = h_r(T_s - T_w) + h_c(T_s - T_a) \tag{1}$$

$$R + C = h(T_s - T_o) \tag{2}$$

Combining (1) and (2) and assuming $h = h_r + h_c$

$$T_o = \frac{h_r T_w + h_c T_a}{h_r + h_c} \tag{3}$$

where T_w = wall temperature, and
h_r and h_c = heat-transfer coefficients for radiation and convection.

STANDARD OPERATIVE TEMPERATURE. Standard operative temperature is a refinement of the operative-temperature analogue to include air velocity. For unclothed subjects at rest in a semireclining posture, Gagge found $T_{so} = 0.48T_w + 0.19[\sqrt{V}T_a - (\sqrt{V} - 2.76)T_s]$.

COOLING POWER. Cooling power is a measure of the atmosphere as a heat sink. A katathermometer is an example of such a device.

EFFECTIVE TEMPERATURE SCALE. Effective temperature scale was established from dry- and wet-bulb temperatures with human comfort as an end point. The scale is specific for the activity and clothing conditions under which it is derived.

WET-BULB GLOBE TEMPERATURE (WBGT). WBGT is an index which may substitute for the effective temperature.

$$\text{WBGT} = 0.2T_g + 0.1T_a + 0.7T_{wb}$$

where T_g = the temperature of a blackened copper sphere 6 in. in diameter,
T_a = air temperature, and
T_{wb} = wet-bulb temperature

THE CLO UNIT

A further simplification and approximation was proposed by Gagge, Burton, and Bazett, who introduced some general terms for metabolism (met) and insulation of clothing (clo). Again, considering only nonevaporative heat loss, we obtain

$$R + C = \frac{1}{I_a + I_{cl}}(T_s - T_a)$$

where I_a and I_{cl} = insulation of air and clothing.

In experiments with resting subjects sitting comfortably indoors, with T_a of 21°C and humidity less than 50 per cent and air movement of 10 cm/sec, they found M to be 50 kcal/m^2 · hr and estimated $R + C$ to be 38 kcal/m^2 · hr; T_s was 33 °C. Thus;

$$I_a + I_{cl} = \frac{T_s - T_a}{R + C} = \frac{33 - 21}{38} = 0.32$$

$$R + C = 0.76M$$

Assigning 0.14 to I_a, they defined I_{cl} as 0.18°C/kcal · m^2 · hr. They also defined the met unit as 50 kcal/m^2 · hr.

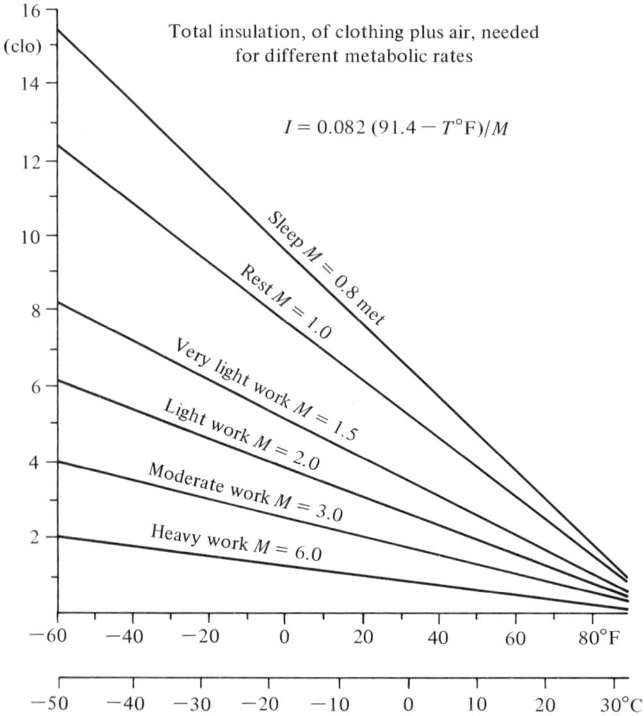

FIGURE 4-3. Total insulation ($I_{cl} + I_a$) for different environmental temperatures and metabolic rates. For the insulation of clothing required (I_{cl}), the insulation of air (I_a), dependent on wind, must be subtracted from the total insulation. (From A. C. Burton and O. G. Edholm. *Man in a Cold Environment.* London: Edward Arnold, 1954.)

It is important to note that, as defined by Gagge et al. 1 clo unit is that amount of thermal insulation that will maintain a resting, sitting man whose metabolism is 50 kcal/(m^2 • hr) indefinitely comfortable in an environment of 21°C, relative humidity less than 50 per cent, and air movement of 10 cm/sec. The number of clos needed to maintain comfort will therefore depend on activity (number of mets) and air movement (which will tend to reduce I_a). Figure 4-3 illustrates this point.

Air movement serves to reduce I_a:

$$R + C = \frac{T_s - T_a}{I_{cl} + I_a - W}$$

where $W =$ wind decrement.

By rearrangement, we obtain

$$R + C = \frac{T_s - [T_a - W(R + C)]}{I_{cl} + I_a}$$

Since $R + C = 0.76M$, increased activity will have the effect of increasing wind decrement for temperature. Figure 4-4 illustrates the thermal wind decrement for various wind speeds at three levels of metabolism. The figure also shows that the decrement is not linear with wind velocity,* and winds above 15 mph have little effect. The equation also shows that as I_{cl} increases, the significance of wind decrement decreases. With standard clothing value, the total insulation $(I_{cl} + I_a)$ could change by 45 per cent as wind increased $(I_a - W = 0)$, whereas with four times the standard clothing this change would be only 16 per cent.

To continue the analysis of environmental stress as it has been presented by Burton, the final consideration at air temperature below 21°C involves the heat gain from solar radiation (R). This heat will be added at the clothing surface:

$$H_s = \frac{T_s - T_{cl}}{I_{cl}}$$

$$H_s + R = \frac{T_{cl} - T_a}{I_a}$$

$$H_s = \frac{T_s - (T_a + RI_a)}{I_{cl} + I_a}$$

* $W = I_a \dfrac{I_a \beta \sqrt{V/V_o}}{1 + I_a \beta \sqrt{V/V_o}}$

where $\beta = 12$ kcal/m^2 · hr · °C, V = m/sec.

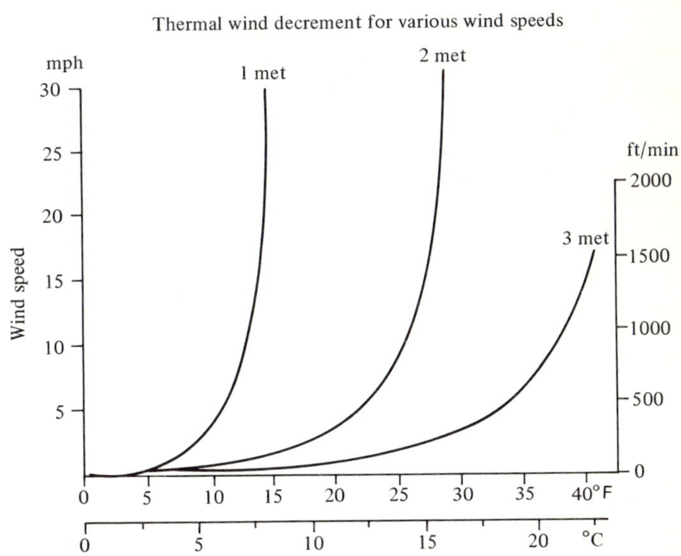

FIGURE 4-4. The thermal wind decrement to be subtracted from the thermometer reading to give the equivalent still-air temperature. (From A. C. Burton and O. G. Edholm. *Man in a Cold Environment.* London: Edward Arnold, 1954.)

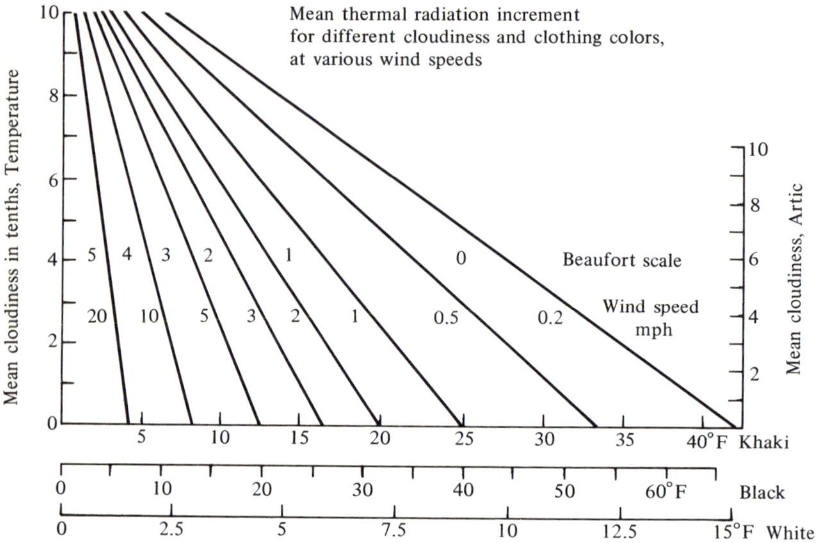

FIGURE 4-5. Chart for estimating thermal radiation increment to be added to the temperature to give the equivalent shade temperature. (From A. C. Burton and O. G. Edholm. *Man in a Cold Environment.* London: Edward Arnold, 1954.)

Figure 4-5 illustrates the radiation increment of different cloudiness, clothing colors, and wind speeds. As I_{cl} increases, the addition from radiation will decrease in significance.

QUANTITATIVE VALUES IN HEAT EXCHANGE

Some examples of changes in energy balance in environments of different temperature in man and animals will serve to illustrate the relevance of consideration of physical factors in thermophysiology.

Classical experiments of Hardy and DuBois (12) and Winslow, Herrington, and Gagge in the United States, and Wezler (19) in Germany have established values for man. One representation illustrates the relation of the various pathways of heat loss after 1 hr of exposure to various environmental temperatures (Figure 4-6). In these experiments, evaporative loss is quantitatively balancing heat gain from metabolism and the environment at the higher temperatures. At low temperature an obvious imbalance occurs, which results in tissue cooling. The heat from tissue cooling is labeled storage. The extent to which tissues cool is illustrated in the second presentation of data (Figure 4-7), which shows the temperature of points on the surface of the body at various environmental temperatures (8). Mean skin temperatures vary with environmental temperature and exposure time. From 30 min to 2 hr are required to reach apparent equilibrium values (19).

Tissue-to-Skin System

Thermal comfort is primarily related to surface temperatures, although internal temperature adds to the subjective feeling. Table 4-1 gives the relative area, the idea temperature, and the regional heat loss at the ideal temperature expressed for the actual area and on the basis of a unit area. Skin temperatures are measured to be representative of these body areas in the human,

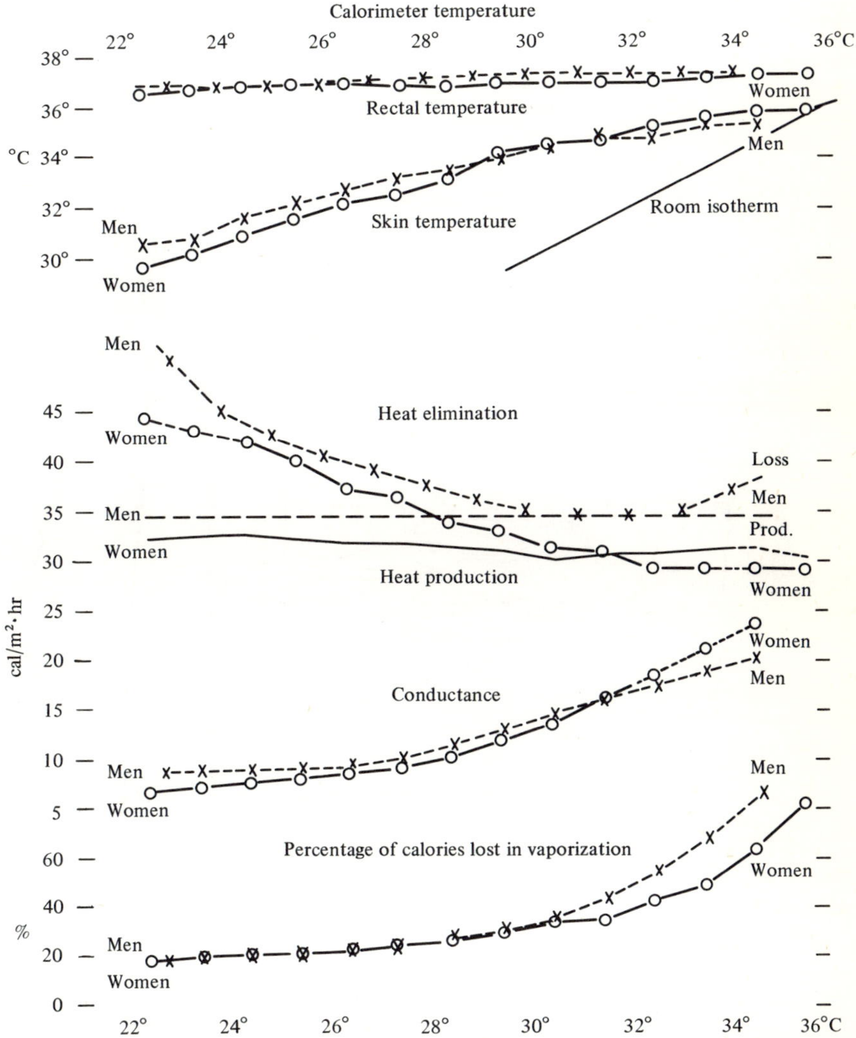

FIGURE 4-6. Comparison of smoothed curves of heat exchange data for 13 women and 2 men. Only the first experimental hour has been used. (From J. D. Hardy and E. F. DuBois. *J. Nutr.* **15:** 477–497, 1938.)

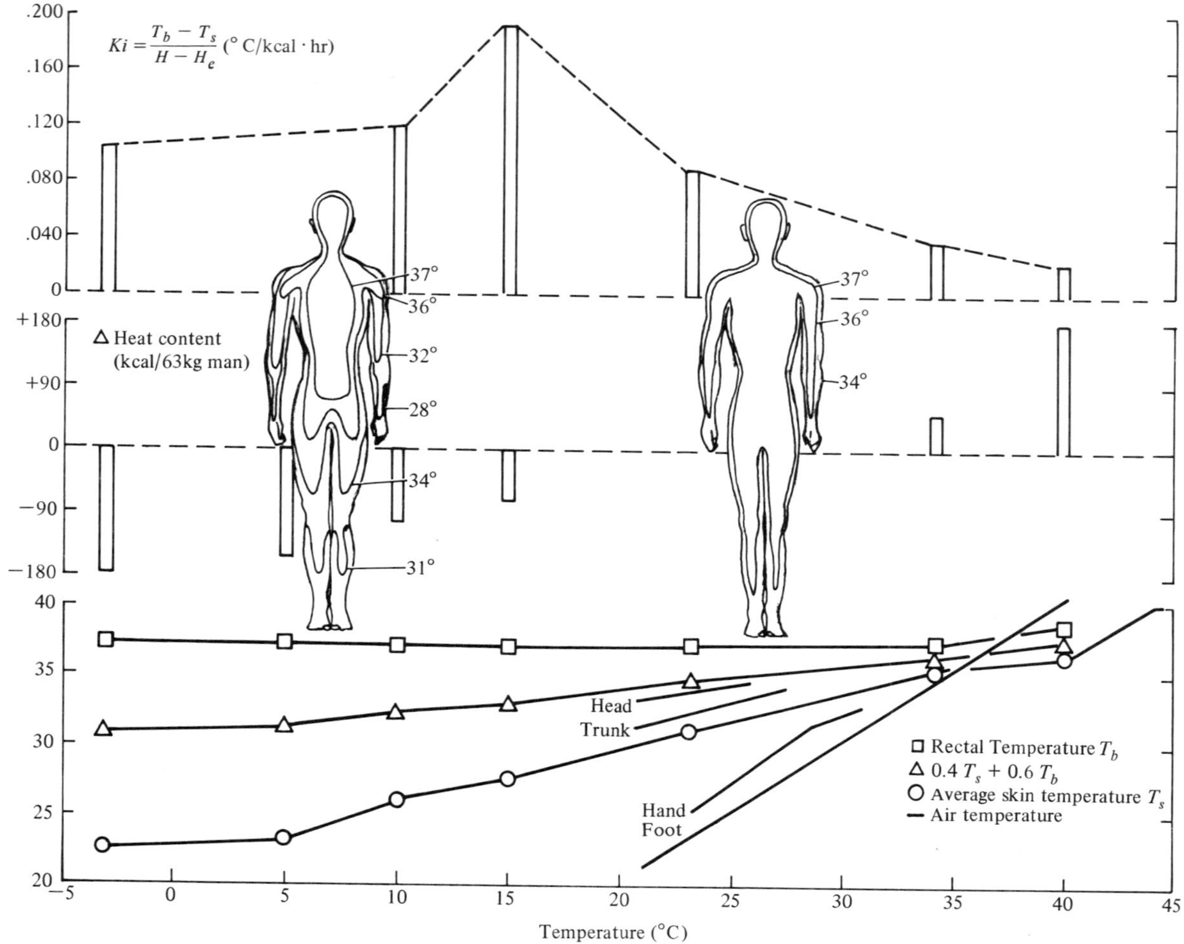

FIGURE 4-7. Tissue insulation *Ki*, change in heat content, and body temperatures related to environmental temperature. (Adapted from L. D. Carlson and A. C. L. Hsieh. Cold. In: *The Physiology of Human Survival.* Edited by O. G. Edholm and A. L. Bacharach. New York: Academic Press, 1965.)

and the relative surface area is a weighting factor for calculating the average skin temperature.

Necessarily, gradients exist between the area of central temperature and the surface. The temperature is normally highest in the central region and decreases as one progresses toward the periphery. The central volume of highest temperature includes the brain and viscera and a variable mass of tissue depending on the environmental temperature. This core is distinguished from a shell in which temperature diminishes toward the periphery. At very warm temperatures the core would extend very nearly to the surface. As the body cools in a cooler environment, heat is lost from tissues as they cool and the shell increases. Figure 4-7 illustrates this concept with two schematic figures. The change in entropy for a 63-kg man has been calculated and the tissue insulation is calculated. Paradoxically, the tissue circulation insulation (inverse of conduction) increases to a maximum and then decreases. At maximum *Ki*, circulation to the periphery is minimal and metabolism is not increased. Below this point, metabolism increases and circulation readjusts to meet metabolic demands.

CONDUCTION AND CONVECTION

Among the many demands made on the circulatory system is the requirement for transport of heat from the site of its production to the periphery for dissipation—and, conversely, for conservation of heat when necessary by reduced flow to the periphery and by countercurrent exchanging. The principal elements in this system are diagramed in Figure 4-8. Blood is uniquely suited to be a heat-transporting system. (Its specific heat is 0.92 cal/gm · °C.) The capillary structure and the A-V anastomoses provide a perfusion possibility that can reduce the conduction distances to the surface to insignificant dimensions. When blood is shunted away from the surface, the insulation of the tissues is interposed to conserve heat.

In the schematic representation shown in Figure 4-8 the pulmonary circulation is shown delivering blood to the lungs. Because the nasal passages, trachea, and bronchi bring the air temperature to body temperature, little heat is lost from the pulmonary circuit. The heat lost as a consequence of ventilation of the lungs is primarily evaporative and is lost from the systemic circulation.

The major source of heat at rest is from the brain and the viscera (72 per cent). (At 50 kcal/m^2 · hr a man of 1.8 m^2 would produce 65 kcal/hr in the brain and viscera and 25 kcal/m^2 in the rest of the body.) The circulation serves to cool these organs and to conduct heat to the periphery, warming peripheral tissues and bringing heat to the surface to be dissipated. In work the muscles may provide 75 per cent of the heat. At a 500-kcal output, 385 kcal would be produced in muscle. At rest, a perceptible change in temperature (0.30° to 0.4°C) would be expected in blood flowing through the brain.

TABLE 4-1
Regional Skin Temperatures and Heat Flows for Thermal Comfort at Rest

Region	Area m^2	Ideal temp. °C	Heat loss at ideal temp. kcal/hr	Heat loss at ideal temp. kcal/m^2·hr	Conductance* kcal/m^2·hr·°C
Head	0.20	34.6	4.0	20.0	7.55
Chest	0.17	34.6	8.2	48.3	18.20
Abdomen	0.12	34.6	4.5	37.5	14.16
Back	0.23	34.6	12.4	53.9	20.35
Buttocks	0.18	34.6	8.3	46.2	17.40
Thighs	0.33	33.0	12.0	36.0	8.55
Calves	0.20	30.8	14.6	73.0	11.30
Feet	0.12	28.6	10.0	83.3	9.65
Arms	0.10	33.0	8.4	84.0	19.80
Forearms	0.08	30.8	8.6	107.5	16.70
Hands	0.07	28.6	16.0	228.6	26.40
Total body	1.80	33.0 (mean)	107.0	59.4	13.98

*Conductance = heat loss at ideal temperature/(ideal temperature − deep body temperature), where deep body temperature is assumed to be 37.25° C.

Adapted from D. R. Burton and L. Collier, The development of water conditioned suits. Royal Aircraft Establishment Technical Note, No. Mech. Eng. 400. London: Ministry of Aviation, 1964. (Unpublished contribution of Dr. McK. Kerslake, Royal Air Force Institute of Aviation.)

In work, the muscle temperature increases, reflecting the inadequacy of the circulation to maintain temperature.

The amount of heat transported from the interior of the body H to its surface is the amount of heat conducted through the tissues plus the amount of heat brought to the surface by the circulating blood:

$$H = \frac{h_{kb}}{d_b}(T_b - T_s) + Fc(T_{ar} - T_{ve}) \tag{1}$$

where h_{kb} = transfer coefficient for conductive heat from the interior to skin,
F = blood flow to the skin, or the periphery,
c = specific heat of blood, and
d_b = the distance between the points at which T_b and T_s are measured.

On exposure to cold the body may conserve heat by reducing F to a minimum. T_s falls as a result of the reduced blood flow but h_{kb}/d_b is also reduced because of the increase in the "shell" of the body; that is, the distance between T_b and T_s is increased. As the blood is diverted to deeper tissues more tissues are interposed between T_b and T_s. The more subcutaneous fat one has, the greater the increment of d_b when blood is diverted to deep tissues. The reciprocal of h_{kb}/d_b is insulation I_b. There is thus an inverse relationship between blood flow to the surface and I_b. When blood flow is minimum, I_b is maximum. I_b has been found to be inversely related to body density and directly related to lipid in the surface tissues. This relationship between I_b, on the one hand, and subcutaneous fat and blood flow, on the other, is often overlooked in the simplified equations describing blood flow to the skin.

When the body is in thermal equilibrium that is, S = 0,

$$M - E = H = K + C + R$$

Assuming $T_{ar} = T_b$ and $T_{ve} = T_s$:

$$\left(Fc + \frac{h_{kb}}{d_b}\right)(T_b - T_s) = h_{kcrs}(T_s - T_a) \qquad (2)$$

where h_{kcrs} = heat-transfer coefficient for K, C, and R.

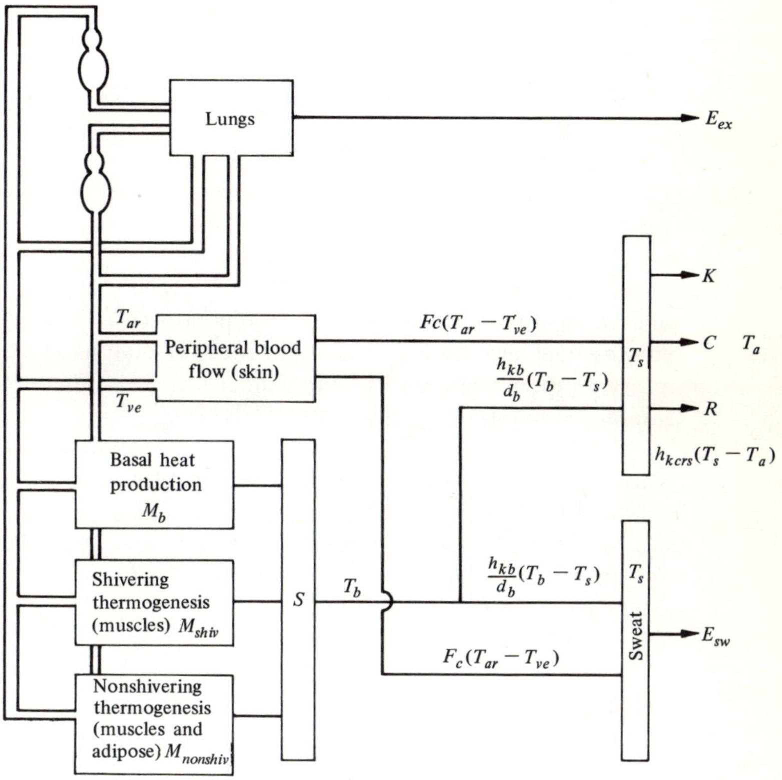

$$M \pm S = E \pm K \pm C \pm R$$
$$M = M_b + M_{shiv} + M_{nonshiv}$$
$$E = E_{ex} + E_{sw}$$

FIGURE 4-8. Model of heat conduction and convection from interior of the body to skin.

$$Fc = h_{kcrs} \frac{(T_s - T_a)}{(T_b - T_s)} - \frac{h_{kb}}{d_b} \qquad (3)$$

Ignoring h_{kb}/d_b and assuming c and h_{kcrs} to be constant, F will vary with the ratio $(T_s - T_a)/(T_b - T_s)$. This is Burton's *thermal circulation index.*

Hardy and Soderstrom (13) found h_{kb}/d_b to be 9.1 kcal/m^2 · hr · °C when the average d_b was 2 cm. Again, assuming $T_{ar} = T_b$ and $T_{ve} = T_s$, equation (1) can be rewritten:

$$Fc = \left[\frac{H}{(T_b - T_s)} - 9.1\right]\frac{1{,}000}{60} \qquad (4)$$

$$F = 17\left[\frac{H}{(T_b - T_s)} - 9.1\right] \qquad (5)$$

where F = blood flow in ml/m^2 · min,
$c = 1$ kcal/l · °C, and
$H - M - E$.

Note that 1,000/60 is used to convert l/hr to ml/min.

It should be noted that Burton's and Hardy and Soderstrom's equations hold only for steady-state conditions.

Another way of describing the variations in temperature in the body is to examine intravascular and tissue temperatures. Data from man in a cooling environment are shown in Figure 4-9A. Rectal temperature and femoral artery temperature are maintained during the exposure while thigh temperature and femoral vein temperature fall (10).

Similar measurements, shown in Figure 4-9B, illustrate the effect of warming and subsequent cooling on temperatures in the femoral artery and vein and in the thigh muscle and subcutaneous tissue. During the period of warming, vein temperature exceeds artery temperature. The increasing femoral artery temperature may indicate some heat is being exchanged between artery and vein (9).

COUNTERCURRENT HEAT EXCHANGE

Bazett (3) suggested that the peripheral vessels in man were arranged to participate in heat exchanging when the periphery was cooler or warmer than the central part of the body. Scholander (16, 17), Aschoff (1, 2), and Schmidt-Nielsen (15) have documented this fact for animals.

Fluids of different temperatures flowing in opposite directions (that is, countercurrent) through contiguous tubes exchange heat across the walls that separate them. The heat flow from one fluid to the other will depend on

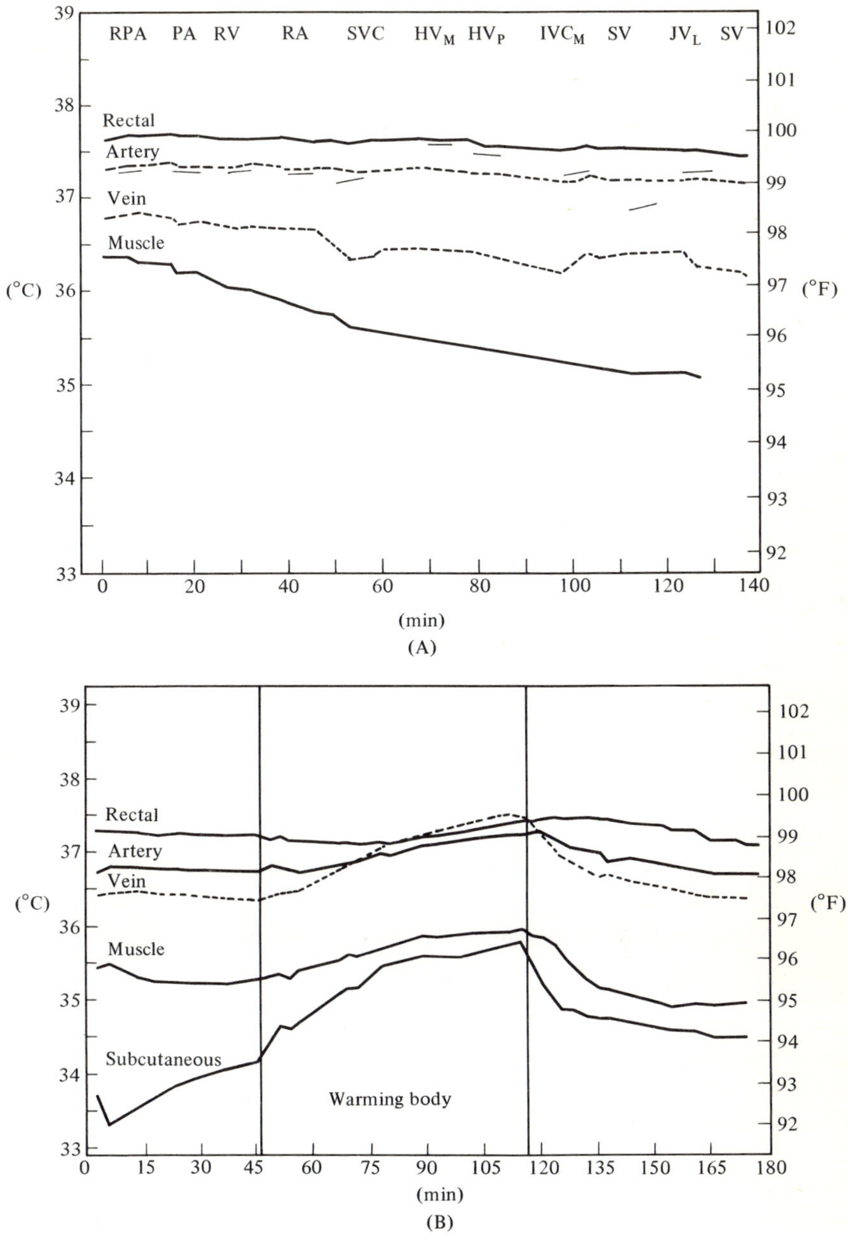

FIGURE 4-9. A: The lines from above downward indicate, in turn, rectal, femoral arterial, femoral venous, and intramuscular (thigh) temperature. The heavy, discontinuous solid bars indicate the temperature obtained with the catheter thermocouple in the locations indicated by the letters at the top. RPA: right pulmonary artery; PA: main stem of pulmonary artery; RV: right ventricle, RA: right atrium; SVC and IVCm: superior vena cava and inferior vena cava (middle); HV_m and HV_p: hepatic vein midliner and hepatic vein periphery; JV_L: left jugular vein; SV: subclavian. (From L. W. Eichna *et al*. Comparison of intracardiac and intravascular temperature with rectal temperature in man. *J. Clin. Invest.* **30:** 353–359, 1951.)

B: Relationship of the changes in rectal, femoral arterial, femoral venous, intramuscular and subcutaneous temperatures during warming and subsequent cooling of the body. Heating of the body was accomplished with blankets and heating pads. (From L. W. Eichna. Thermal gradients in man. *Arch. Phys. Med.* **30:** 584–593, 1949.)

1. The temperature difference between them (ΔT).
2. The surface available for heat transfer (A).
3. The mass flow and specific heats of the fluids (F.S.ρ).
4. The conductance of the thermal barrier between the two fluids (k).

The conductance k sums all the factors affecting heat flow per unit area per unit temperature difference—for example, the conductivity and thickness of the tissue separating the two fluids, the boundary conditions within the tubes, and the character of the fluid flow.

The heat transferred from the warmer to the cooler fluid can be represented as H where

$$H = k \cdot \text{F.S.}\ \rho \cdot A \cdot \Delta T$$

Detailed considerations of countercurrent heat exchangers and the calculations that relate to them can be found in most textbooks dealing with heat transfer.

Both Scholander and Krog (16) and Tregear (18) offer brief discussions of the main characteristics of countercurrent heat exchange systems. According to Scholander and Krog.

$$F(\Delta T + \Delta t) = QL\ \Delta t + \Delta t F$$

where ΔT = temperature drop across the heat exchanger,
Δt = temperature difference between the two fluids,
F = the mass flow,
Q = the conductance of the thermal barriers, and
L = the exchange area.

That is, the amount of heat entering the exchanger is equal to the heat dissipation into the medium surrounding the exchanger plus the countercurrent exchange.

The major features of single countercurrent heat-exchanging situations are illustrated by Scholander and Krog's model (Figure 4-10).

1. The heat exchanged per unit length depends on the mass flow—that is, the higher the flow, the smaller the temperature difference.
2. The *temperature difference* between the fluids is constant along the exchanger for a given mass flow.

Examination of the vascular system in many regions of most mammals and birds shows two veins accompanying each artery, not only in the limbs but also beneath the dermis, generally, and in the viscera. Such an arrangement suggests that heat transfers between the blood flowing toward the periphery and the blood returning from it in circumstances in which there

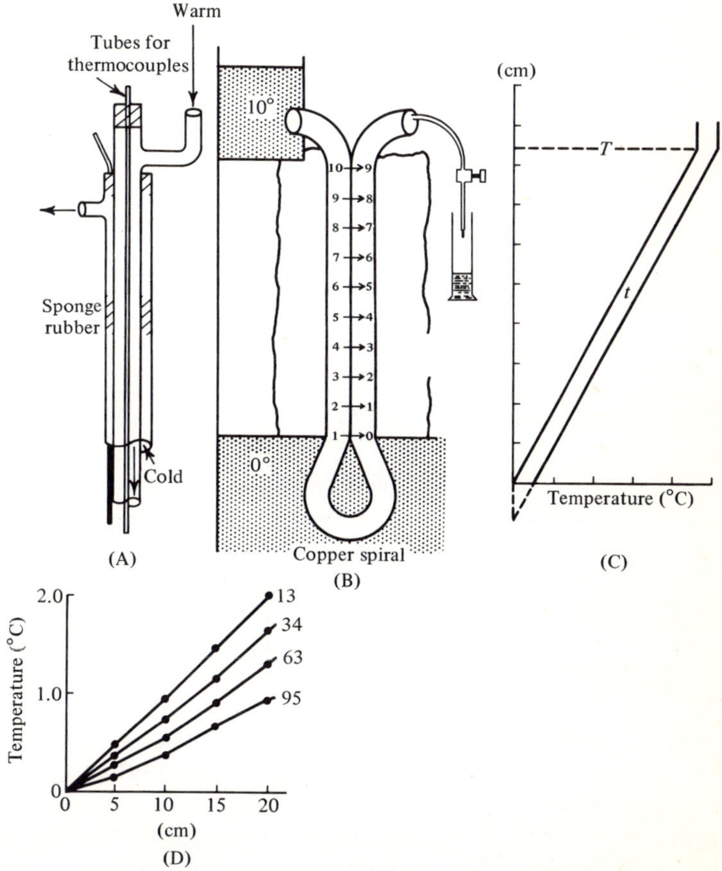

FIGURE 4-10. A: Upper part of concentric countercurrent cooler used. B: Temperature gradients in the model. C: Graphic representation of temperature relations. D: Temperature gradients in cooler at flow rates of 13, 34, 63 and 95 ml/min. (Adapted from P. F. Scholander and J. Krog. Counter-current heat exchange and vascular bundles in sloths. *J. Appl. Physiol.* **10:** 405–411, 1957.)

is a temperature difference between the two streams (Figure 4-11). Bazett and his coworkers (3) have demonstrated the presence of longitudinal temperature gradients within the arteries and their venae comitantes in the human arm and leg. These gradients are normally 0.03°C/cm but, in cold conditions, can increase to 0.35°C/cm. In the latter case, the arterial blood entering the hand was about 21°C. Bazett's data and the subsequent studies of Scholander and others suggest an important role for arteriovenous countercurrent heat transfer in decreasing the conductance along the length of limbs.

Scholander has related the specialized vascular arrangements in Arctic mammals and birds, in marine mammals, and in slow-moving animals, such as the sloth and anteater. The remarkable feature of the vascular systems of

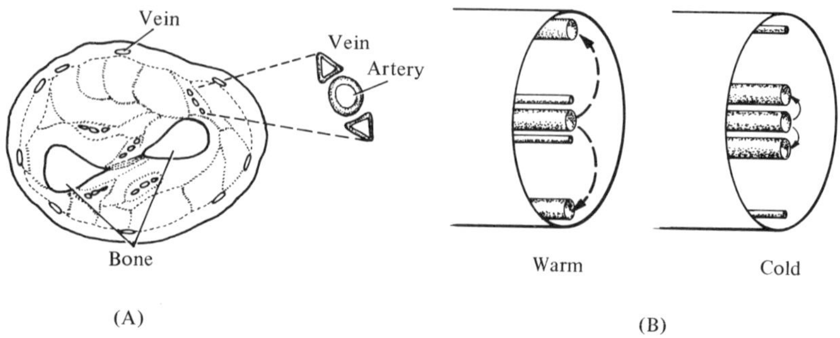

FIGURE 4-11. Heat exchange system in the forearm. A: is a diagram of a cross section of the forearm showing the superficial and deep veins and their relationship to arteries. B: is a diagrammatic representation of the shift in venous blood flow from superficial vessels in warm surroundings to deep-lying vessels in the cold.

these animals is the subdivision of large arteries into numerous branch arteries, each accompanied by two veins. The most remarkable of these "radiator" arterial rete arrangements has been described by Fawcett (11) in the manatee (sea cow). In this animal the more usual arboreal branching of arteries is replaced in many regions by a branchlike structure with many small arteries, each terminating in a capillary bed. Each artery is accompanied by veins.

In many of the regions in which countercurrent heat exchange has been either demonstrated (by measurement of intravascular temperatures) or inferred from surface temperature distributions and vascular morphology, the venous drainage provides the possibility of a superficial return path separated by considerable thicknesses of tissue from the arteries and a deep path via the veins that accompany the arteries. These two pathways are illustrated in Figure 4-11. Although the potential for thermoregulatory control of heat storage and peripheral conductance has been suggested by several people, so far there is no real information relating to control of venous return pathway.

Countercurrent heat transfer within the vascular system has been suggested as the explanation of several characteristically different physiologic phenomena. Some of the suggested functions of these arrangements are as follows:

1. A means of decreasing the temperature gradient along the length of limbs and so conserving body heat while permitting a reasonably high nutritive blood supply to the extremities.
2. As with the testis, helping to preserve a testicular temperature favorable to gametogenesis by precooling of the blood flowing through the spermatic artery by heat transfer from arterial blood to the veins of the pampiniform plexus that return blood from the *superficial* veins of the scrotum.

Comparative Thermophysiology

The preceding examples deal with human responses to different environmental temperatures. These stresses are dealt with in a variety of ways by different animals. The comparative thermophysiology of homeotherms, as well as the special examples in hibernating animals, elucidate differing mechanisms of meeting the environmental stress. A special difference in the area of evaporative loss exists between animals that pant and those that sweat.

The classic example of Gelineo, illustrated in Figure 4-12, indicates the difference in metabolic response to temperature of animals adapted to various environments. This concept has been expanded by Scholander and co-

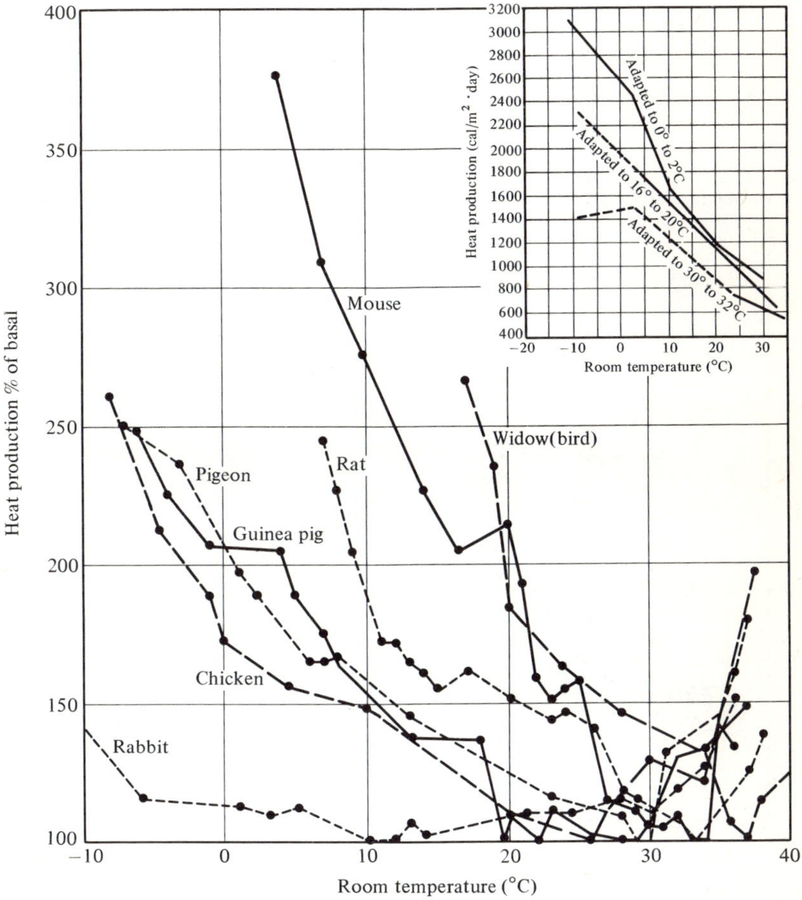

FIGURE 4-12. Influences of environmental temperature on heat production of animals expressed as percentages of the minimum or basal heat production. (From S. Brody. *Bioenergetics and Growth.* New York: Reinhold, 1945. By permission of Van Nostrand Reinhold Company.)

workers (17). Their representation of insulation with respect to fur thickness, given in Figure 4-13, shows the animal kingdom in comparison and illustrates the difference between air and water.

Animals also illustrate the extent to which surface temperature may vary on the body and extremities (Figure 4-14). A specific example in the illustration of the profile of a Merino sheep shows the heat balance and the interaction of radiation and convection. In this example, the following data indicate the input-output relation.

Input		**Output**	
Metabolic heat	40 kcal	Free convection	60 kcal
Conduction from the air	0	Radiation	190
Radiation from the sun	240		250 kcal
Radiation from the sky	60		
Radiation from the ground	50		
	390 kcal		

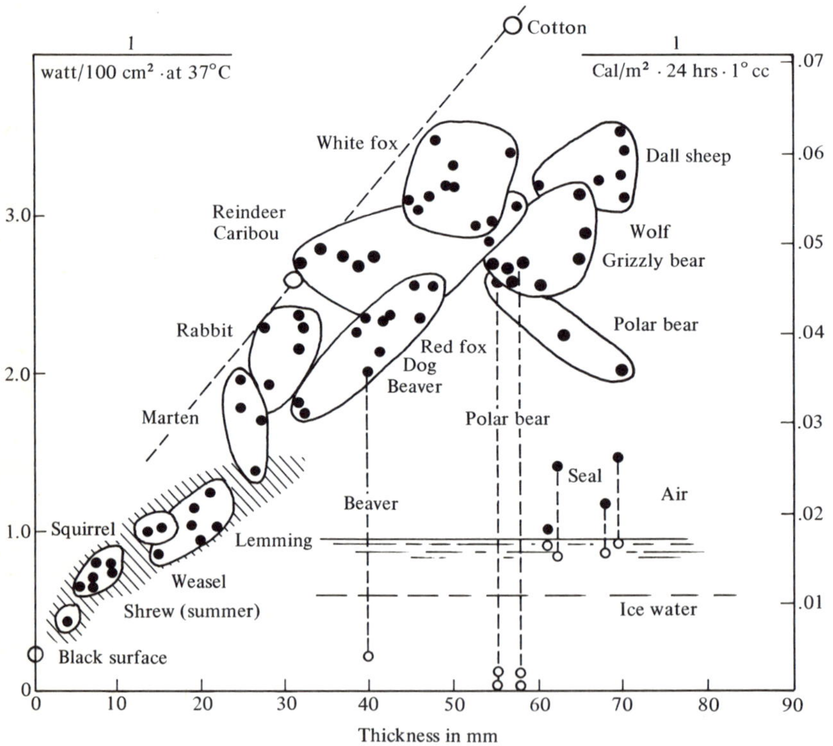

FIGURE 4-13. The relationship of insulation to fur thickness in a series of mammals. The shaded area indicates tropical mammals. In aquatic mammals (seal, beaver, polar bear) the measurements in air are corrected with measurements taken in ice water. (From P. F. Scholander *et al.* Body insulation of some arctic and tropical mammals and birds. *Biol. Bull.* **99:** 225–236, 1950.)

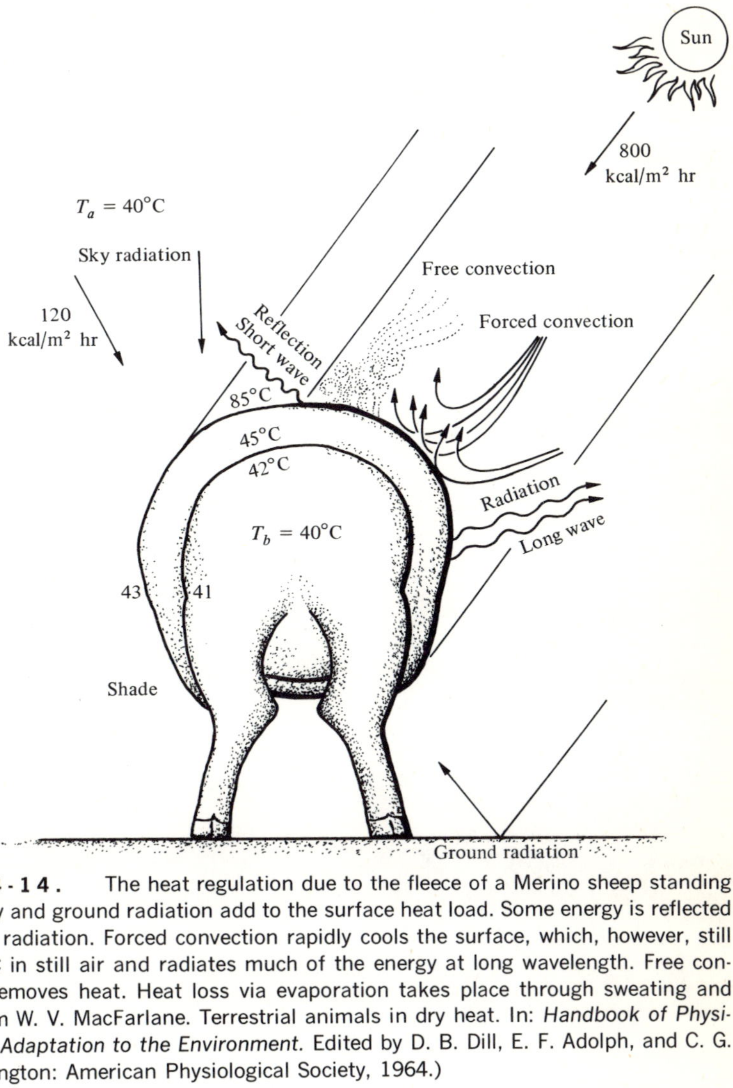

FIGURE 4-14. The heat regulation due to the fleece of a Merino sheep standing in the sun. Sky and ground radiation add to the surface heat load. Some energy is reflected as short-wave radiation. Forced convection rapidly cools the surface, which, however, still heats at 85°C in still air and radiates much of the energy at long wavelength. Free convection also removes heat. Heat loss via evaporation takes place through sweating and panting. (From W. V. MacFarlane. Terrestrial animals in dry heat. In: *Handbook of Physiology,* Sec. 4, *Adaptation to the Environment.* Edited by D. B. Dill, E. F. Adolph, and C. G. Wilbur. Washington: American Physiological Society, 1964.)

Assuming body weight to be 50 kg and specific heat to be 0.8, this animal would warm at the rate of 3.5°C per hour were it not for evaporative loss. Air movement would also remove the heat from the surface of the wool. At night the radiation loss to the sky might be as great as 200 kcal, with additional loss to the ground. The behavioral effects on this exchange, such as the animal lying down or exposing minimal area to the sun, are easily imagined.

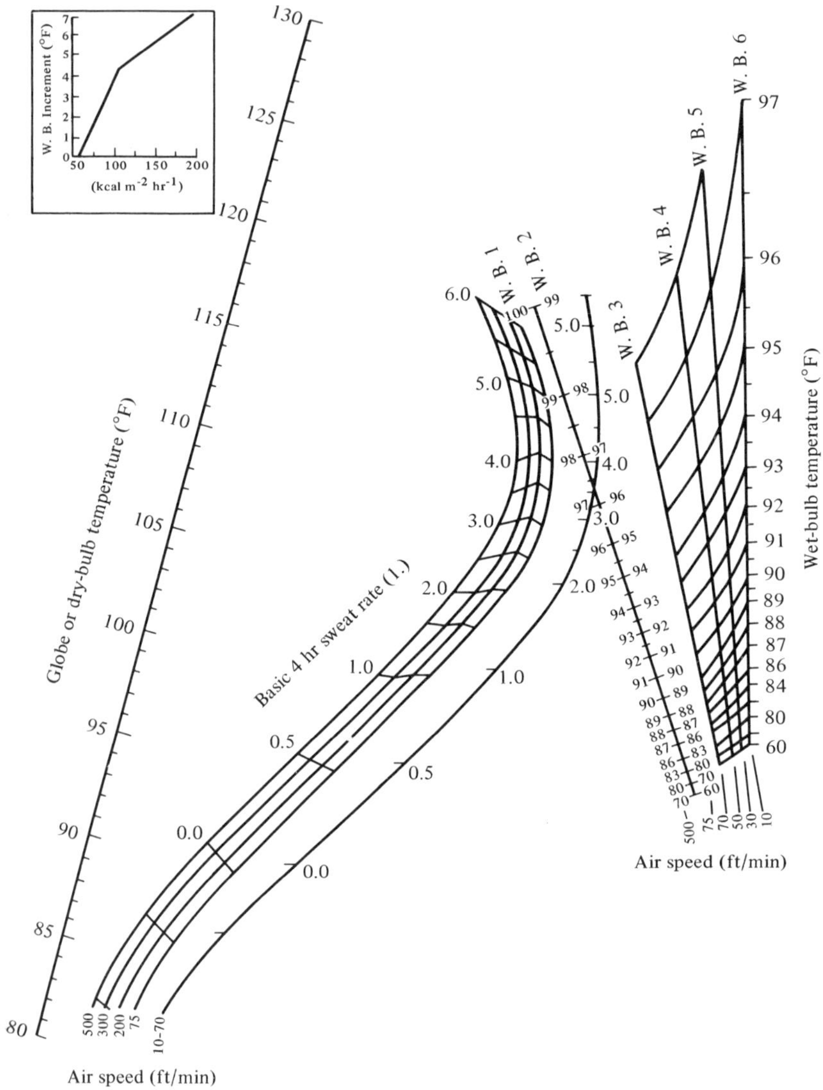

FIGURE 4-15. The predicted 4 hr sweat rate index (P_4SR). (From R. K. Macpherson. Physiological response to hot environment. Special Report Series No. 298, London: Her Majesty's Stationery Office, 1960. Printed by permission of the Controller of Her Majesty's Stationery Office.)

Predictions and Evaluation of Stress

The preceding discussion has been directed toward the evaluation of the physical factors important in heat loss. A number of scales have been derived to integrate these values into a single value.

PROCEDURE

Enter A, dropping vertical line from intercept of globe temperature with air speed; gives combined heat load of radiation and convection. Extend vertical line to

Enter B. At intercept with metabolism, draw horizontal line; obtain total heat load in terms of evaporation required for heat balance (E_{req}). Extend horizontal line to *Enter C.*

Enter X. Draw horizontal line from intercept of dry- and wet-bulb temperature; obtain vapor pressure gradient between saturated skin at 95°F and ambient air. Extend horizontal line to

Enter Y. At intercept with air speed draw vertical line; obtain maximum evaporation from wet skin at 95°F (E_{max}). Extend vertical line to

Enter Z. Proceed to intercept with horizontal line from C. (If E_{max} exceeds 2400, enter Z at 2400).

Read heat stress index value.

EX: Metabolism 600 (light arm work standing at bench). For solution follow broken lines: Heat stress = 90.

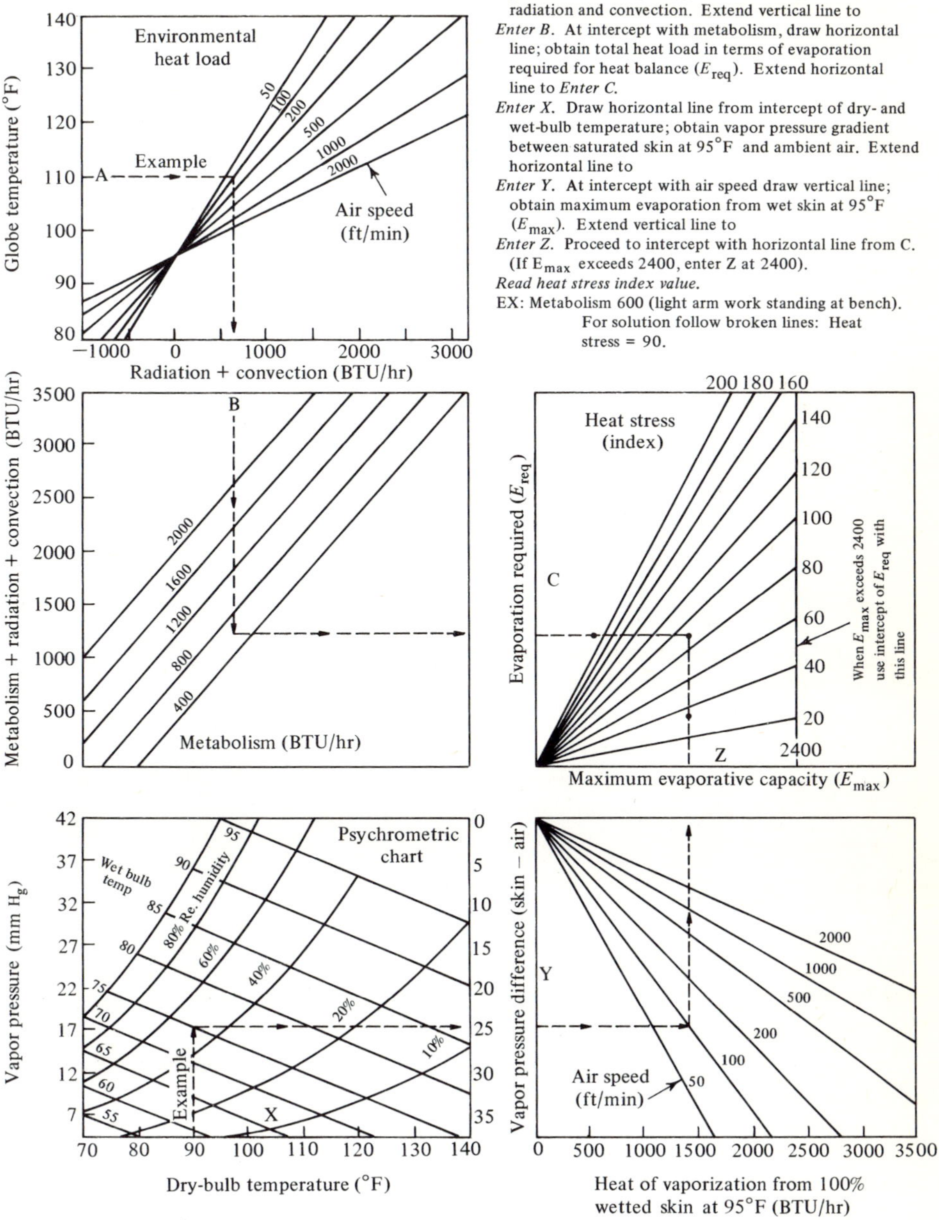

This index represents a rational approach to evaluation of heat stress in terms of: 1. thermal load which is imposed on a standard man; 2. capacity of the environment to accept the load; and 3. physiological capacity to meet the demands over a period of 8 hours.

The standard man represents a composite of young, fit, acclimatized men who have been subjects of physiological investigation of the effects of heat in the laboratory and in the field.

FIGURE 4-16. Flow charts for determining heat stress index values. Example: globe 110, dry bulb 90, wet bulb 75, air speed 100, metabolism 600. For solution, follow broken lines: Heat stress = 90. (From H. S. Belding and T. F. Hatch. Index for evaluating heat stress in terms of resulting physiological strain. *Heating, Piping and Air Conditioning* **27:** 129, 1955.)

For cold the Wind Chill Index was devised. This is based on a prediction of the rate at which the naked body would cool in subfreezing temperatures and varying wind velocities. The empirical equation

$$K_o = (\sqrt{V \times 100} + 10.45 - V)(33 - T_a)$$

where K_o = cooling power of the atmosphere in kcal/hr · m^2,
V = wind velocity in m/sec, and
T_a = temperature of air in °C.

A wind-chill scale varying from 0 to 2,600 was calibrated against subject performance (18). This criterion is used by the U. S. Department of the Army for prediction of cold stress and a similar value is published in some Canadian newspapers.

For heat a predicted 4-hr sweat rate has been devised from the results of a large series of experiments. Sweat production is a good indication of the physiologic strain imposed on the individual. A P_4SR nomogram is given in Figure 4-15.

Another predictor of load (Belding-Hatch index) represents an approach in terms of thermal load imposed on a man, the capacity of the environment to accept the load, and the physiologic capacity to meet the demands over a period of 8 hr. (See Figure 4-16.) These nomograms average values for young men and serve to provide criteria to engineers and estimates to industrial medical officers concerning working conditions. The nomograms also indicate the prime variables in thermal stress (4).

References

1. Aschoff, J. Warmeanstausch in eiver modellextremitat. Pfluger. Arch. Ges. *Physiol.* **26:** 260–271, 1957.
2. Aschoff, J., and R. Wever. Kern und schale in warmhaushalt des menschen. *Naturwissenschaften* **45:** 477, 1958.
3. Bazett, H. C., L. Love, M. Newton, L. Eisenberg, R. Day, and R. Forster, II. Temperature changes in blood flowing in arteries and veins in man. *J. Appl. Physiol.* **1:** 3–19, 1948.
4. Belding, H. S., and T. F. Hatch. Index for evaluating heat stress in terms of resulting physiological strain. *Heating, Piping and Air Conditioning,* **27:** 129, 1955.
5. Burton, A. C. The application of the theory of heat flow to the study of energy metabolism. *J. Nutr.* **7:** 497–533, 1934.
6. Burton, A. C. The average temperature of tissues of the body. *J. Nutr.* **9:** 261–280, 1935.

7. Burton, A. C., and H. C. Bazett. A study of the average temperature of the tissues, of the exchange of heat and vasomotor responses in man by means of a bath calorimeter. *Amer. J. Physiol.* **117:** 36–54, 1936.
8. Carlson, L. D., and A. C. L. Hsieh. Cold. In: *The Physiology of Human Survival.* Edited by O. G. Edholm and A. L. Bacharach. New York: Academic Press, 1965, Chap. 2.
9. Eichna, L. W. Thermal gradients in man. *Arch. Phys. Med.* **30:** 584–593, 1949.
10. Eichna, L. W., A. R. Berger, B. Rader, and W. H. Becker. Comparison of intracardiac and intravascular temperature with rectal temperature in man. *J. Clin. Invest.* **30:** 353–359, 1951.
11. Fawcett, D. W. A comparative study of blood vascular bundles in the Florida Manatee and in certain Cetaceans and Adentates. *J. Morphol.* **71:** 105–124, 1942.
12. Hardy, J. D., and E. F. DuBois. Basal metabolism, radiation, convection and vaporization at temperatures of 22 to 35°C. *J. Nutr.* **15:** 477–497, 1938.
13. Hardy, J. D., and G. F. Soderstrom. Heat loss from the nude body and peripheral blood flow at temperatures of 22°C to 35°C. *J. Nutr.* **16:** 493–510, 1938.
14. Honda, N., L. D. Carlson, and W. V. Judy. Skin temperature and blood flow in the rabbit ear. *Amer. J. Physiol.* **204:** 615–618, 1963.
15. Schmidt-Nielsen, K. In: *Temperature, Its Measurement and Control in Science and Industry.* New York: Reinhold, 1963, Vol. III.
16. Scholander, P. F., and J. Krog. Counter-current heat exchange and vascular bundles in sloths. *J. Appl. Physiol.* **10:** 405–411, 1957.
17. Scholander, P. F., V. Walters, R. Hock, and L. Irving. Body insulation of some arctic and tropical mammals and birds. *Biol. Bull.* **99:** 225–236, 1950.
18. Siple, P. A. Measurements of dry atmospheric cooling in sub-freezing temperature. *Proc. Amer. Phil. Soc.* **89:** 177–199, 1945.
19. Tregear, R. T. In: *Physical Functions of Skin.* New York: Academic Press, 1966.
20. Wezler, K. Physiological fundamentals of hyperthermia and pathological physiology of heat injury. In: *German Aviation Medicine, World War II.* Washington, D. C.: U. S. Govt. Printing Office, 1950, Vol. II, Chap. VIII D.

Additional Reading

Brody, S. *Bioenergetics and Growth.* New York: Reinhold, 1945.

Burton, A., and O. G. Edholm. *Man in a Cold Environment.* London: Edward Arnold, 1954.

Irving, L. Terrestrial animals in cold. In: *Handbook of Physiology,* Sec. 4. *Adaptation to the Environment.* Edited by D. B. Dill, E. F. Adolph, and C. G. Wilbur. Washington: American Physiological Society, 1964.

Leithead, C. S., and A. R. Lind. *Heat Stress and Heat Disorders.* Philadelphia: F. A. Davis, 1964.

MacFarlane, W. V. Terrestrial animals in dry heat. In: *Handbook of Physiology,* Sec. 4, *Adaptation to the Environment.* Edited by D. B. Dill, E. F. Adolph, and C. G. Wilbur. Washington: American Physiological Society, 1964.

Newburgh, L. H. *Physiology of Heat Regulation.* Philadelphia: Saunders, 1949.

CHAPTER

5

Regulation of Temperature

The Concept of Normal Temperature

Mammals are homeotherms. This term implies that the body temperature is kept within certain limits. Just as our ideas of overall metabolism tend to reflect a standard rate of metabolism to all tissues, there is an implicit understanding that there is a standard, or "normal," body temperature. The range of deep body temperature of homeotherms of different species is small, extending from 36°C in the elephant to 41°C in birds, in spite of the very wide range in the temperature of their natural habitat. This has led to some interesting speculation on how the "normal" temperature evolved. Burton and Edholm, assuming that development of homeothermic animals took place in areas where the mean annual temperature is about 25°C, suggest that the range of body temperatures actually found is a compromise between two disadvantageous temperature ranges. A body temperature regulated at a low temperature would mean that the excess temperature would be small and changes of environmental temperature would call for relatively greater proportional changes in the physiologic regulatory mechanisms. In the language of the environmental physiologist, the same amount of stress would result in a greater strain on the animal. (See introduction to Chapter 6.) On the other hand, there is a good deal of evidence of a lethal temperature at about 43°C. The range from 36 to 41°C would represent a useful compromise.

The Body Temperature

The term *body temperature* is a misnomer. It is merely the temperature of deep central areas including the heart, lungs, abdominal organs, and brain. The traditionally measured and compared temperatures of the body are oral, rectal, and, in some cases, axial.

The variations of temperature within the body, as well as at the surface, have been mentioned and the point has been made that the temperature of various tissues is dependent on blood flow to the tissues. In some instances, blood flow is a means of cooling the tissue; in other instances, the blood flow acts as a means of keeping the tissue warm. It is clear that there will be deviations in deep body, or core, temperature depending on the activity of an individual. Indeed, there are variations in certain areas which follow the rhythm of respiration caused by the alterations in blood flow which follow respiratory patterns.

The normal distribution of body temperature, measured at a specific time of day, is illustrated in Figure 5-1. There are wider variations imposed on the usual range occasioned by age, activity, excitement, and environmental temperature (Figure 5-2).

There is a clear circadian rhythm in temperature. The extent of this rhythm is indicated in Figure 5-3. The work of Aschoff (3) has demonstrated that this temperature is normally locked into a 24-hr cycle by *zeitgebers*. If such *zeitgebers* as light-darkness cycles are not present, the temperature pattern develops into a free-running rhythm usually slightly longer than 24 hrs in duration.

In the female a temperature pattern follows the menstrual cycle and, indeed, a rather sharp change in temperature associated with ovulation has been used as such an index (Figure 5-4).

Basic Regulatory Mechanisms

From the over-all equation for energy exchange given in Chapter 4, and the subsequent discussion, the following equation can be derived:

$$T_b = I(M - E) + T_a$$

where I = insulation extending from T_b to T_a, and
T_a = a composite temperature of the environment.

In order to maintain T_b constant in the face of fluctuations in T_a, the body has at its disposal means of altering I (insulation), M (metabolism), and E (evaporative heat loss). Alterations in I and E are usually referred to as physical regulation; changes in M are called chemical regulation. Because

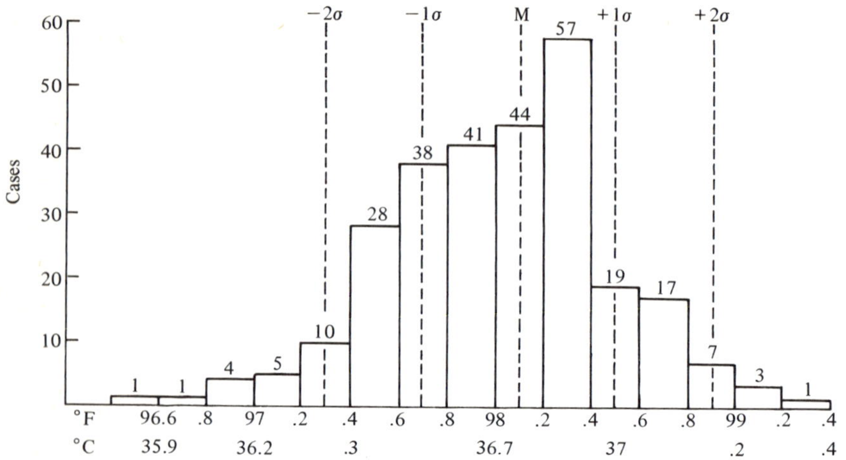

FIGURE 5-1. Histogram of oral temperatures of 276 medical students seated in class between 8 and 9 A.M. Mean, 98.1 ± 0.4°F. (From A. C. Ivy. What is normal or normality? *Quart. Bull. Northwestern Univer. Med. Sch.* **18:** 22–28, 1944.)

variations in *I* would involve less expenditure of resources (energy and body fluids), there is a teleologic reason why the body should rely on this mechanism first. In addition to these physiologic changes, the animal may resort to certain overt behavioral actions, such as change in posture and movement.

Only the neural aspects of temperature regulation will be discussed, but it should be kept in mind that the responses are modified by endocrine and behavioral factors. The response of the system will depend on input signals

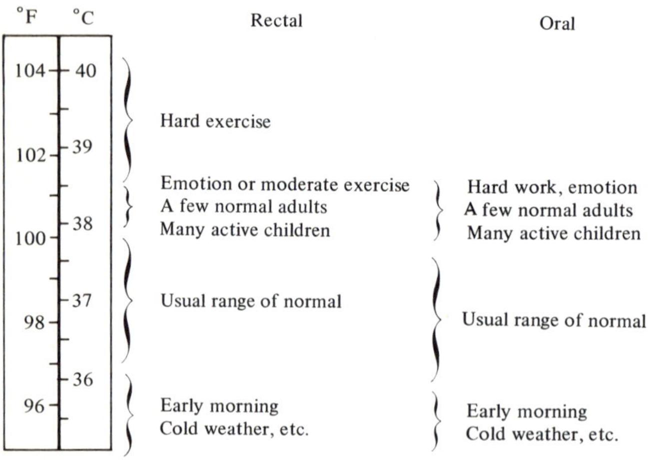

FIGURE 5-2. An estimate of the ranges in body temperatures found in normal persons. (From E. F. Dubois. *Fever and the Regulation of Body Temperature.* Springfield: Thomas, 1948.)

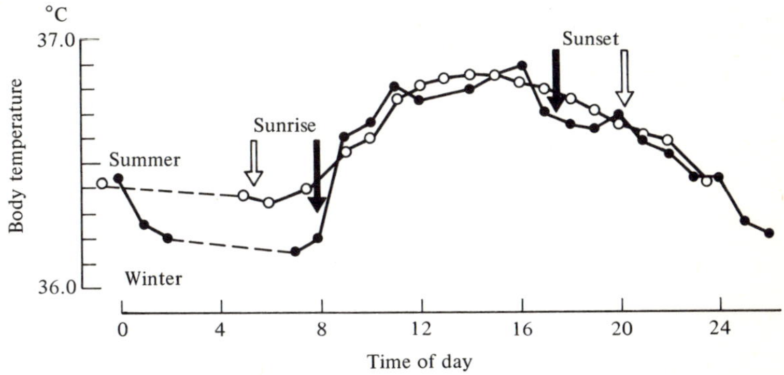

FIGURE 5-3. Circadian rhythm in body temperature. (From T. Sasaki. Effect of rapid transposition around the earth on diurnal variation in body temperature. *Proc. Soc. Exp. Biol. and Med.* **115:** 1129–1131, 1964.)

reaching a central integrating area from which output signals effect the change in insulation (blood flow or piloerection) or heat production (shivering or nonshivering thermogenesis). It will be clear that the power of the system, as reflected in the size of the response and the ability of the animal to maintain body temperature, will be a function of all the factors modifying the neural signals as well as of those that determine the ability of the tissues to respond.

Peripheral Sensory System

The neural responses to a negative heat load depend on the input from cold receptors and a sensing system for the central temperature and perhaps for the body heat content. The peripheral sensors are histologically defined. The cold receptors are situated close to the surface of the skin; their number varies with the particular area of the skin. The number is much higher in the skin of the face and hands than it is on the legs or chest. The numbers per unit area are illustrated in the schematic diagram shown in Figure 5-5.

The characteristics of cold receptors have been carefully studied by Hensel (18), who used the classic neurophysiologic technique of single-unit analysis. The salient features of these investigations are illustrated in Figure 5-6. The frequency of discharge is related to the temperature and to the rate of change in temperature. A sharp change in temperature causes a large increase in frequency, the magnitude of which depends on the extent of the temperature change. If the change of temperature occurs over a long period, the burst frequency is less. Whether the change in temperature is rapid or slow, the new temperature gives a different discharge frequency. When the temperature is raised, the cold receptors become silent and then return to the frequency characteristic of the new temperature. Individual cold receptors have a range of temperature to which they respond; the range varies according to the fiber.

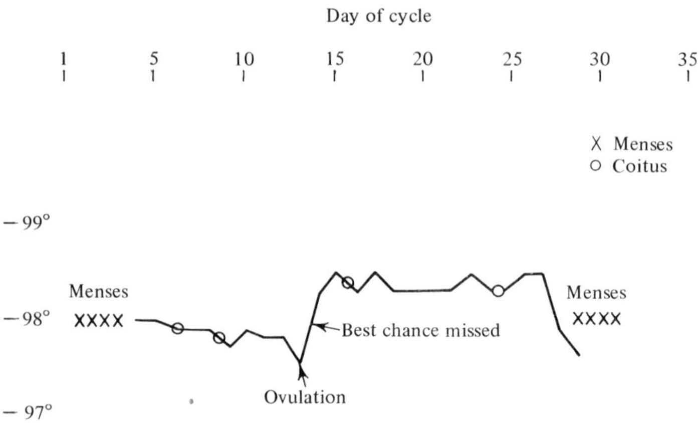

Basal temperature record of a typical menstrual cycle

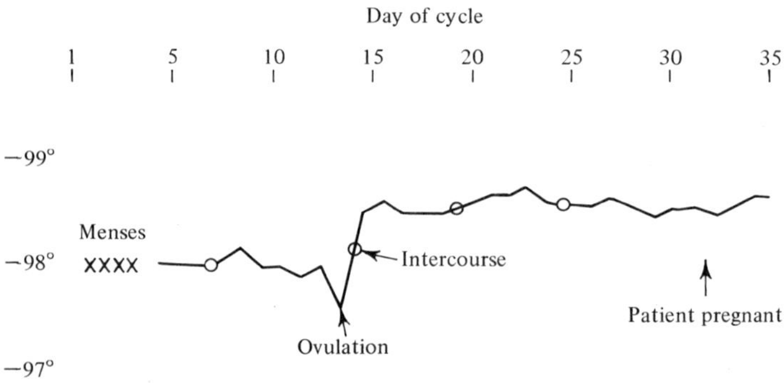

Basal temperature record of a cycle followed by pregnancy

FIGURE 5-4. Body temperature, recorded daily before arising, of a woman during two menstrual cycles. The second cycle is followed by pregnancy. (From W. A. Selle, *Body Temperature: Its Changes with Environment, Disease and Therapy.* Springfield: Thomas, 1952.)

The normal temperature-sensing range of the skin is represented by distribution curve for these fibers. The nerve fibers that originate from cold receptors, together with the pain fibers, are predominantly in the lateral spinothalamic tract, where they form synaptic connections and project to the somatesthetic area. These pathways from the periphery can be followed to the thalamus. Because the center for control of temperature has been defined within the hypothalamic area, connecting pathways must exist to bring this input into evidence for the regulating center (Figure 5-5).

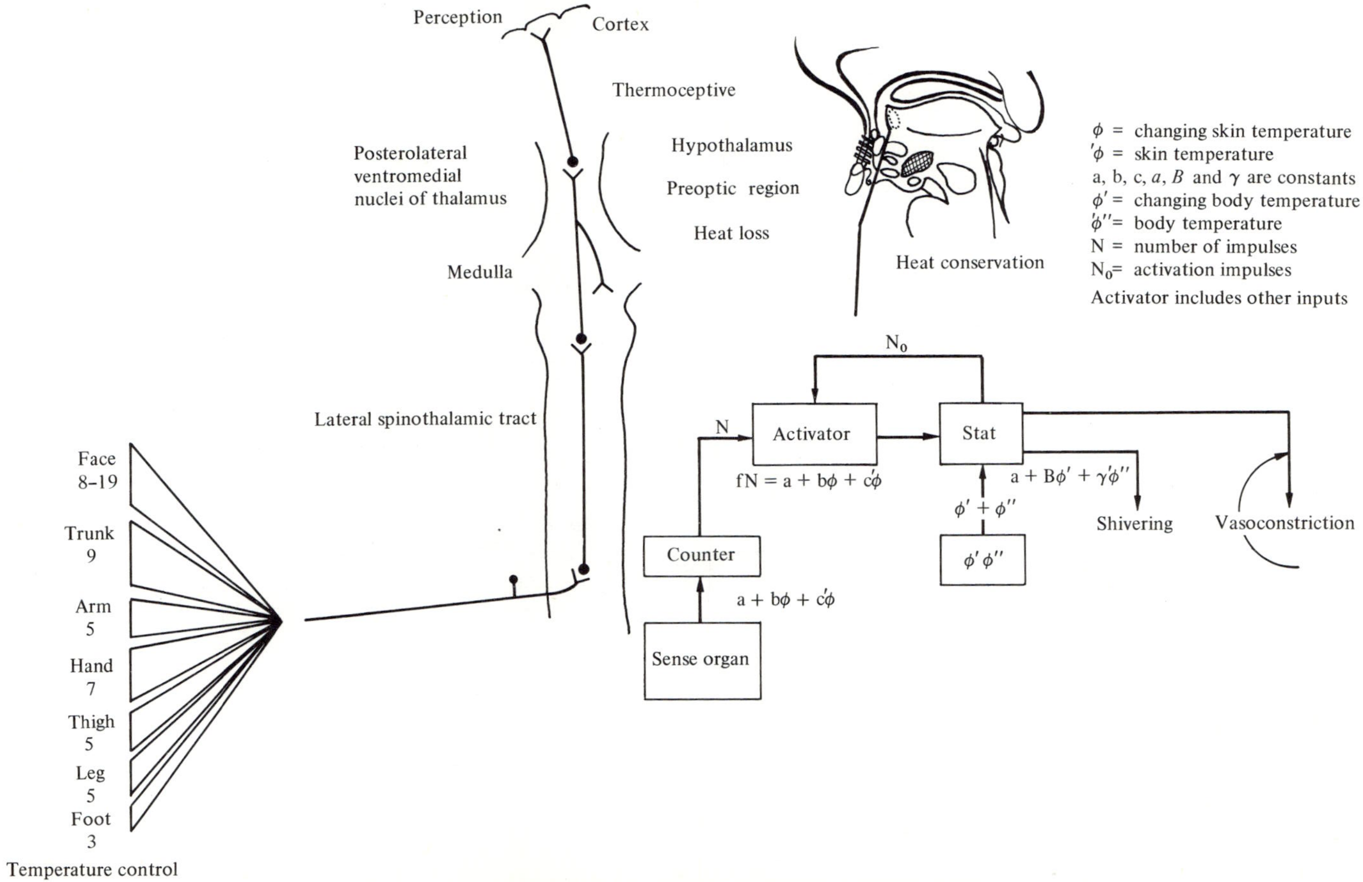

FIGURE 5-5. The sensory input for temperature regulation. In addition to peripheral receptors, a thermoceptive area in the hypothalamus has been postulated.

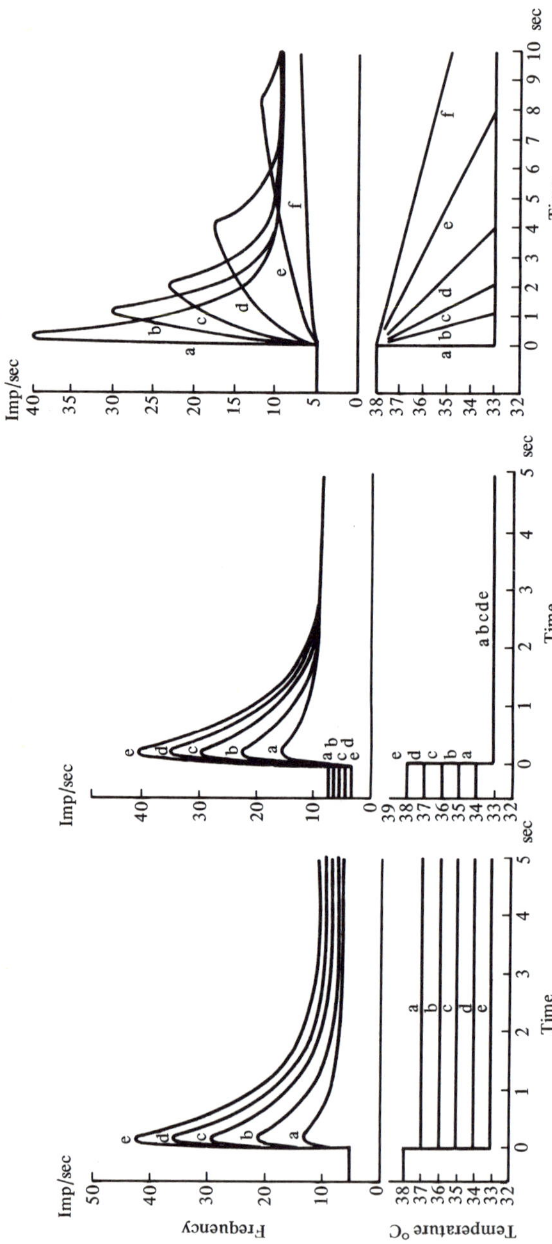

FIGURE 5-6. Impulse frequency of single units related to temperature change. (From H. Hensel. Physiologie der thermoreception. *Ergebn. Physiol.* **47:** 166–368, 1952.)

Central Sensory and Integrating System

THE HYPOTHALAMUS

Efforts to localize areas in the brain that are involved in temperature regulation involve classic lesion procedures, electrical stimulation and recording, and methods of changing brain temperature by perfusion of the brain and local heating and cooling with thermodes. Spontaneous measurements of brain temperature have also been made.

Lesion, stimulation, and recording techniques have established functional areas in the hypothalamus. These studies have been accomplished on dogs, cats, rabbits, and goats. Results of the studies are related to a drawing of the human brain in Figure 5-7 to establish the general areas involved in control.

The classic studies of Keller (19) established the location of heat-regulating centers by determining the approximate amount of tissue that must be destroyed to eliminate an animal's ability to regulate against a cool or warm environment. Heat-loss control is in the anterior hypothalamus. Animals with lesions in the cephalic midbrain (level 1 to 3, Figure 5-7) can regulate their body temperature when placed in a cool environment but do not prevent a body temperature rise when placed in a warm environment. Conversely, a dog with a posterior hypothalamic lesion (level 4 to 6, Figure 5-7) can pant when exposed to heat but fails to vasoconstrict or shiver when it is exposed to a cold environment and its body temperature falls.

Lesion experiments have been extended by Hemingway and coworkers (17) to delineate areas of the brain involved in the shivering response. In cats, lesions in the posterior hypothalamus, the upper midbrain, the lower midbrain, the upper and lower pons, and the medulla and spinal cord abolish or spare shivering. The best definition of the primary areas involved in shivering is the dorsomedial region of the posterior hypothalamus in the dorsal wall of the third ventricle between the mammillary bodies and the tuberal region (level 3 to 6, Figure 5-7). Stimulation experiments confirm this localization. The efferent pathway is not the rubrospinal or periventricular hypothalamic pathways but rather a diffuse descending system merging into the lateral white columns of the spinal cord.

Andersson and colleagues (1, 2) showed similar area involvements in cold-defense reactions in the goat. Electrical stimulations in or in the vicinity of the septum pellucidum of unanesthetized goats were observed to produce shivering, peripheral vasoconstriction, and sometimes piloerection. Andersson and his colleagues were also able to demonstrate that in animals exposed to heat electrical stimulation in this area would inhibit polypneic panting. Andersson, Grant, and Larson continued their investigations to areas that were involved in the central control of heat-loss mechanisms in the goat. They defined the heat-loss center as occupying the dorsal half of the area between the anterior commissure and optic chiasma lateral to the medial edge of the

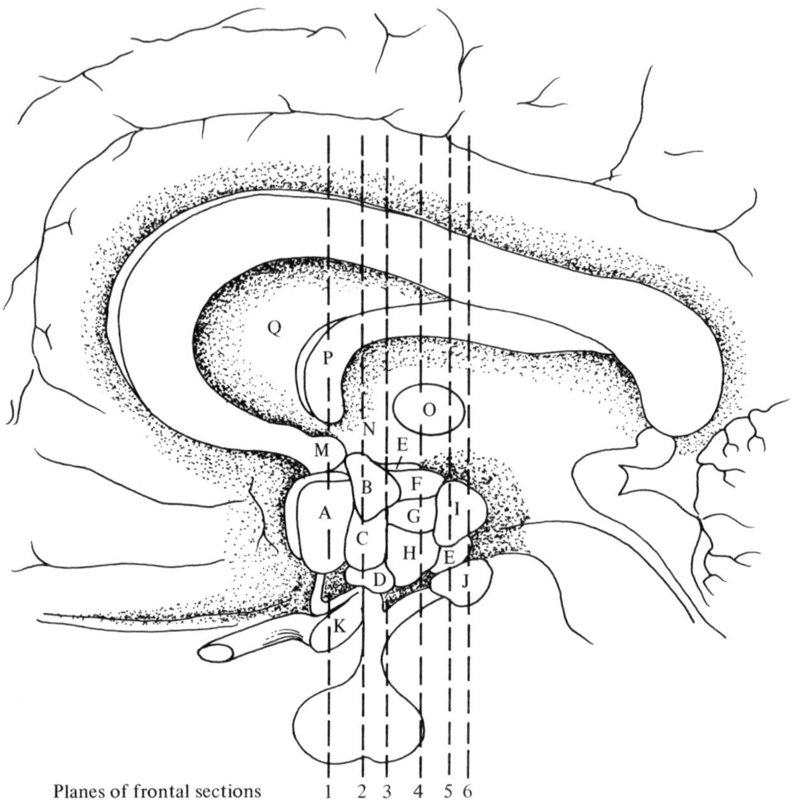

A	Preoptic nuclei
B	Paraventricular nucleus
C	Anterior hypothalamic area
D	Supra-optic nucleus
E	Lateral hypothalamic area
F	Dorsal hypothalamic area
G	Dorsomedial nucleus
H	Ventromedial nucleus
I	Posterior hypothalamic area
J	Mammillary body
K	Optic chiasma
L	Lamina terminalis
M	Anterior commissure
N	Hypothalamic sulcus
O	Intermediate mass of thalamus
P	Fornix
Q	Septum pellucidum

FIGURE 5-7. Sagittal section of human brain with level designations (caudal to rostal) for projected localization of areas found to be associated with temperature regulation in animal experimentation. Lesions must be made bilaterally in order to disturb function.

internal capsule. Experiments with the use of thermodes, regulation of the temperature of the blood perfusing the head, and application of cool or warm solutions in the ventricle have supported the hypothesis.

Thermoceptive units exist in the hypothalamus. These thermoceptive units have been shown by intracellular techniques to respond to changes in hypothalamic temperature with changes in impulse frequency which are different from those of other cells in the brain. Thermoceptive neurones have been found at the level of the anterior commissure rather widely distributed by

Nakayama and his coworkers (22). Warm- and cold-sensitive neurons have also been located in the dog brain by Cunningham and coworkers (12), and these results have been confirmed by Wit and Wang (28).

Specific areas of the hypothalamic area can be shown to be sensitive to leukocytic pyrogens (10). In Figure 5-8 the sagittal section of the rabbit brain, 1.5 mm to the right of the midplane, shows the site of injection of the leukocyte pyrogen and the time between the injection and the onset of fever. If one accepts the location at a point of minimal time difference, the area is just at the level of the optic chiasma.

Andersson and his coworkers have demonstrated that local cooling of the heat-loss center caused a marked thyroid activation indicated by an increase in plasma protein-bound iodine and a fall in the amount of radioactivity in the thyroid gland. There is a time delay in the response of 30 min to several

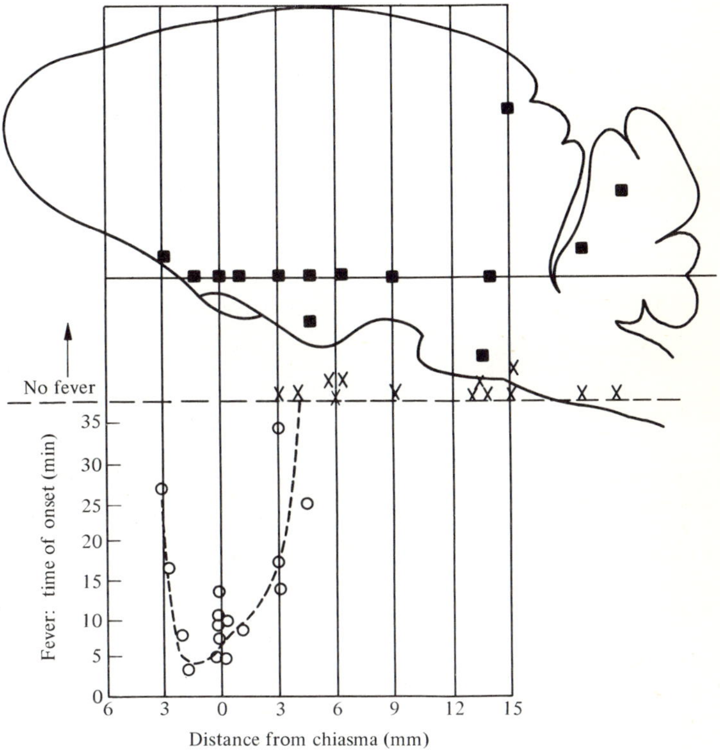

FIGURE 5-8. Sagittal section of rabbit brain 1.5 mm to the right of the midplane. Closed squares indicate site of injection of leucocyte pyrogen; the crosses indicate no fever resulted and the circles indicate the time of onset of fever after injection in minutes plotted against distance from the optic chiasma. (From K. E. Cooper, W. I. Cranston, and A. J. Honour. Observations on the site and mode of action of pyrogens in the rabbit brain. *J. Physiol.* **191:** 325–377, 1967.)

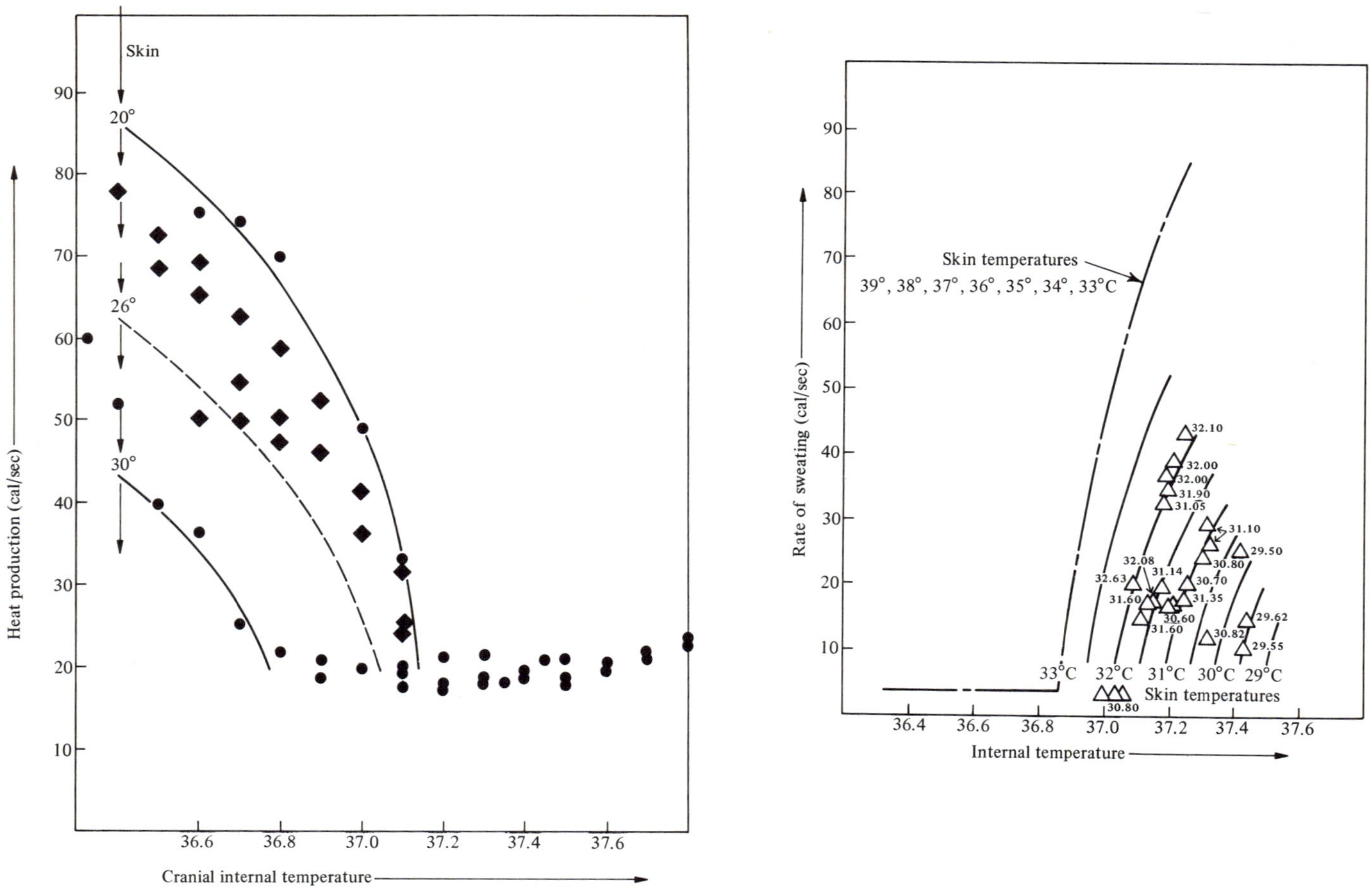

FIGURE 5-9. Results from experiments in the human calorimeter showing that the relationship between heat production or sweat rate and cranial internal temperature is modified by the skin temperature of the subject. (From T. H. Benzinger, C. Kitzinger, and A. W. Pratt. The human thermostat. In: *Temperature: Its Measurement and Control in Science and Industry.* Edited by J. D. Hardy. New York: Reinhold, 1963. Part 3. Chap. 56. By permission of Van Nostrand Reinhold Company.)

hours. It would be reasonable to assume that this is mediated from thyroid-releasing factor and the thyroid-stimulating hormone (TSH). By stimulation of central warm detectors in the preoptic and anterior hypothalamic region an inhibitory effect could be demonstrated on the release of TSH. These experiments also indicate that there may be an effect on the sympathetic-adrenal-medullary system. Local cooling of the preoptic anterior hypothalamic region causes an increase in the excretion of both norepinephrine and epinephrine.

A family of transfer functions may be written relating cranial temperature to heat production, sweat rate, and conductivity (Figure 5-9). The family of functions is directly related to average skin temperature in heat loss and, inversely, in heat conservation. These transfer functions have the characteristics of a threshold which has the properties of a set point, and the temperature-dependent characteristics of the curves are more significantly influenced by the threshold change than by a gain change. Although the afferent input from temperature sensors has been the principal variable in shifting the response parameter, other influences (such as the chemical environment) may influence the response(s).

It seems reasonable to conclude, then, that the hypothalamic area is intimately involved in thermoregulation, not only as a thermoceptive area but also as an area integrating incoming signals from the periphery. The responses mediated by this system include the activation of heat-conservation and heat-loss mechanisms as well as activation of the thyroid and adrenal medulla.

OTHER AREAS

Complete bilateral removal of the sympathetic chains results in a reduction in the animal's ability to maintain temperature at 37°C but not to the extent that the hypothalamic lesions do. Although the concept of a "center" is an attractive one in considering control systems, the term cannot be accepted in a strict sense in temperature regulation because so many organs and systems are involved in the regulation. There are likely to be gradations in regulation resulting from the response patterns of different systems. Thauer and associates (25, 26) suggest a graded system of control along the levels of the nervous system, because cooling the body core elicits shivering even though the brain and cutaneous temperatures are kept at or above normal values. One of the thermal-sensitive core sites appears to be between C_2 and L_7 of the spinal cord in the dog.

The Effector Systems

The effector systems in regulation of temperatures are vasomotor, shivering and nonshivering thermogenesis, and a piloerection in fur-bearing animals for defense against heat loss. As mechanisms for defense against heat gain, vasomotion and sweating or panting are the principal effector systems. In both situations behavioral patterns also serve to select and protect for optimal body temperature.

HEAT PRODUCTION

Shivering pathways involve the motor nerves and their proprioceptive and tension feedback. The graded heat production caused by shivering is dependent on the posterior hypothalamus. The intensity of shivering depends on an interaction between areas in the cortex, anterior hypothalamus, and cerebellum. The rhythmicity is influenced by the cerebellum and the frequency by the interactions in the spinal cord. The oxygen consumption may be increased manyfold by shivering, but the efficiency for maintaining body heat is very low—10 to 15 per cent. Nonshivering thermogenesis is activated by the sympathetic nervous system. The increased heat production may be blocked by autonomic blocking agents and sympathetic blockade. Infused norepinephrine is calorigenic in young animals and in cold-adapted animals. (See discussion of nonshivering thermogenesis in Chapter 6.) The target tissue for the norepinephrine effect which has been implicated is adipose tissue and, most definitely, brown adipose tissue.

BLOOD FLOW

Blood flow in the periphery is mainly distributed to skin and muscle, and these tissues have a rich vasomotor innervation. Some areas—such as the skin of the foot, hand, or ears—are much more actively concerned in temperature regulation than others, such as the trunk or head. Not only is it not possible to extrapolate function from hand skin to arm skin, but the skin of the digits responds differently from the base of the hand. Temperature per se has an effect on the vessels to add to the vasomotor control.

Figure 5-10 illustrates the neural control of blood vessels. Sympathetic nerve constriction of vessels is mediated by norepinephrine to the alpha receptors in the vessel wall. Beta receptor activation leads to dilation. A third receptor is postulated to mediate the effects of metabolic products, and the smooth muscle per se may be influenced by temperature. Central control for heat conservation is localized in the posterior and ventromedial nuclei of the hypothalamus (15).

The blood supply to the skin, especially in the periphery, is richly endowed with A-V anastomoses. Observations on the rate of removal of Na^{24} from the rabbit's foot indicate that all the hyperemia associated with a cold-induced vasodilation and dilation accompanying rewarming following vasoconstriction in the cold is flow through anastomoses.

Body cooling and body heating affect blood flow in the periphery (Figure 5-11). The hand (or foot) blood flow is more sensitive to temperature change

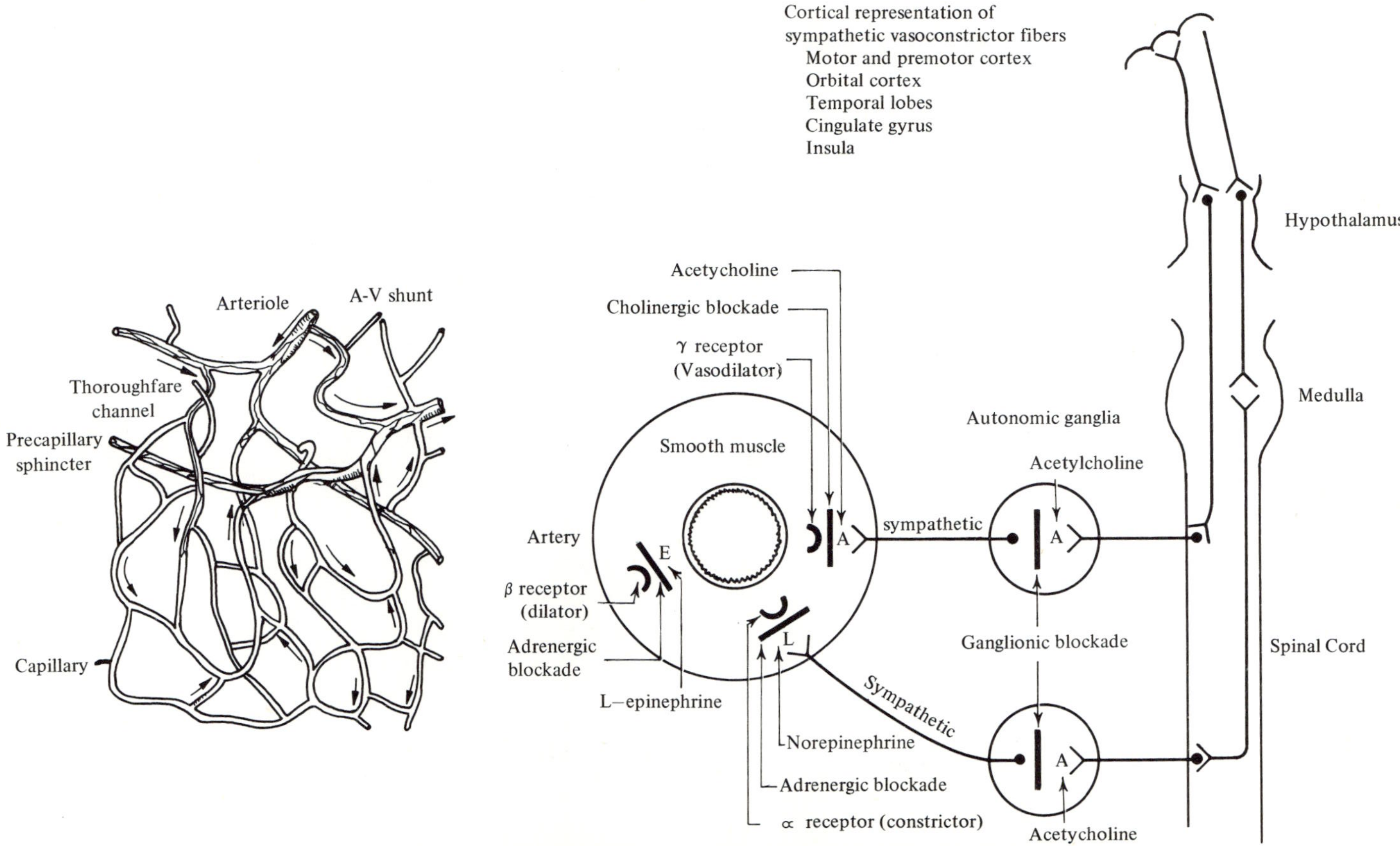

FIGURE 5-10. Schematic diagram of the regulation of arteriolar size (and, hence, bloodflow). On the left is shown an arteriovenous shunting system (A-V shunt) which could direct blood from the capillaries. (Adapted from H. D. Green and J. H. Kepchar. Control of peripheral resistance in major systemic vascular beds. *Physiol. Rev.* **39:** 617–686, 1959.)

Thermal status of subjects	Spealman (1945)	Ferris et al. (1947)	Brown and Page (1952)	Krog et al. (1962)
1. Uncomfortably warm (room temperature)	32°C	30°C	———	25°C
2. Comfortable or cool (room temperature)	24°C	24°C	20°C	———
3. Cold (room temperature)	16°C	17°C	———	———
Hand skin temperature				
Water bath temperature	2 to 35°	———	5° to 45°	2° to 40°
Skin temperature	———	17° to 42°	———	———
Symbols	○	●	x	△

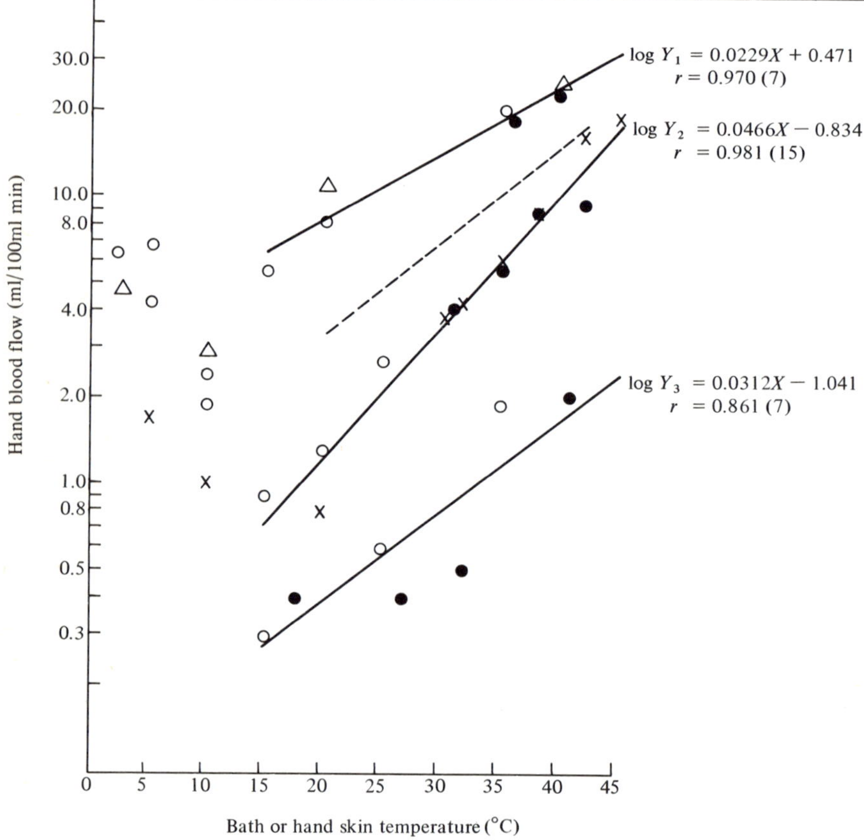

FIGURE 5-11. The relationship between bloodflow through the hand and hand skin temperature, showing the influence of the thermal status of the subjects. (From L. D. Carlson and A. C. L. Hsieh. Cold. In: *The Physiology of Human Survival.* Edited by O. G. Edholm and A. L. Bacharach. New York: Academic Press, 1965, Chap. 2.)

than that of the arm (8). Local temperature also affects blood flow. For example, when the fingers are immersed in water at various temperatures, blood flow decreases from a high value in water at body temperature or above to a minimum at 10 to 15°C. The increase in flow at temperatures below 10°C is characteristically cyclic. The phenomenon known as cold-induced vasodilation was described by Sir Thomas Lewis in 1930 (21). When the fingers are immersed in ice water, the blood flow is quickly reduced to zero and the finger cools to bath temperature. After 7 to 15 min the flow returns and the finger rewarms. If the finger is kept in the cold water, the cycle repeats. The cycle appears to be altered by, but not entirely dependent on, the skin nerve supply (Figure 5-12). The cycling will continue if the skin nerves are blocked. The cycling is suppressed or inhibited if the body core temperature is lowered. It has been postulated that a major cause of cold-induced vasodilation is

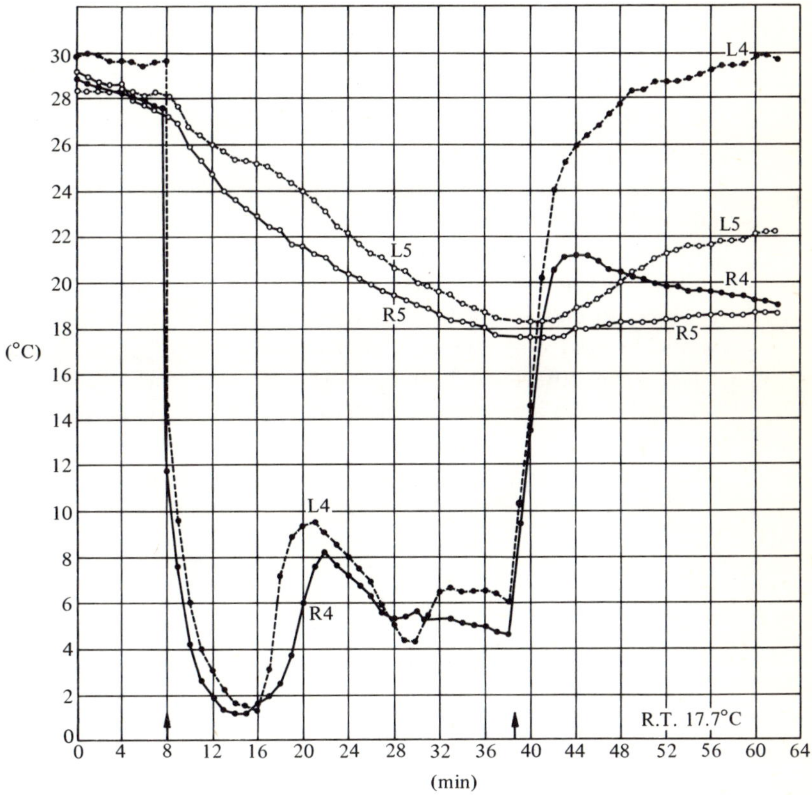

FIGURE 5-12. Skin temperatures from the fifth fingers in room air at 17.7°C and from the fourth fingers before, during, and after immersion in ice water. The rise in temperature on the left (sympathectomized) side appears to be greater. (From T. Lewis. Observations upon the reactions of the vessels of the human skin to cold. *Heart.* **15:** 177–208, 1930.)

impairment of vascular response to constrictor hormones. The possibility that a vasodilator substance is released has not been ruled out. Both the initial blood flow decrease as a response to local cooling and the subsequent cold-induced vasodilation are produced by the coordination of several quite different mechanisms. At least four factors are involved in the initial decrease in flow: the reflex excitation of vasoconstrictor fibers, a direct constrictor response of smooth muscles, the change in blood viscosity, and the local change in metabolism. The cold-induced vasodilation may be due to axon reflexes, release of the local vasoconstriction caused by the low temperature, and a progressive relaxation of the smooth-muscle constrictor response.

When the circulation to the periphery is at its lowest value, the maximum insulation of the tissues will be evident. The reciprocal of this value, conductivity, is given on page 45 as though it were a fixed number. However, the conductivity of a region may vary from 10 to 2.5 kcal/°C·m^2·hr, depending on the amount of body fat (10 to 30 per cent). The calculated conductance* for the human body may increase eightfold because of circulation to the periphery. In the skin of the periphery, such as the hand, the calculated conductance may vary over a much wider range—the blood flow in the finger may vary from 0 to 100 ml/100 ml·finger·min and the heat loss may change by a factor of 1,000 in water.

The relationship of conductance to skin temperature, rectal temperature, and mean body temperature is illustrated in Figure 5-13 at rest and at work. This overall conductance is an average heat-transfer coefficient for conduction and convection, because surface tissues serving as insulation and the circulation to skin and muscle are both involved. A general deduction from Figure 5-13 would be that circulation to skin is the major response to the temperature effect—the difference between the rest and work level is constant when referred to skin temperature. Several mechanisms participate in the regulation of cutaneous blood flow: release of vasoconstrictor tone, vasodilator impulses, and a direct effect at the vessels of a substance, possibly bradykinin, which is released when sweat glands are activated.

SWEAT

Sweat glands are activated cholinergically in man by the sympathetic nervous system. Sweating occurs when the heat load is increased, whether from internal (exercise) or environmental causes. The rate of sweat production is dependent on skin temperature and central body temperature. The classic work of Robinson (23) illustrates this relationship (Figure 5-14). The sweating response of two working subjects referenced to average skin temperature illustrates the threshold and linear relation of sweating to skin temperature.

*Conductance is conductivity modified by other factors, in this case blood flow.

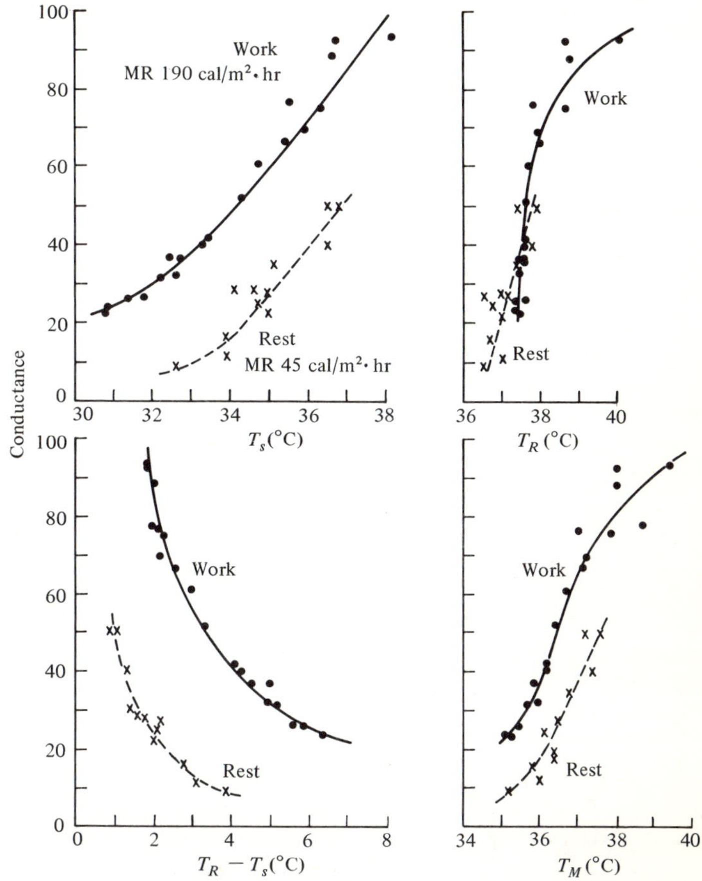

FIGURE 5-13. The effect of work on the relation of tissue heat conductance to body temperature. T_R: rectal temperature; T_S: mean skin temperature; T_M: mean body temperature. (From S. Robinson. Circulatory adjustments of man in hot environments. In: *Temperature: Its Measurement and Control in Science and Industry.* Edited by J. D. Hardy. New York: Reinhold, 1963, Vol. 3, Part 3, Chap. 27. By permission of Van Nostrand Reinhold Company.)

When referenced to rectal temperature, the relationship seems to emphasize the effectiveness of the skin temperature signal in early phases when skin temperature is changing.

Summary

In Figure 5-15, the concepts of temperature regulation are summarized under the headings of the controller and the controlled system. Accepting the premise that a primary neural integrating area exists in the hypothalamus,

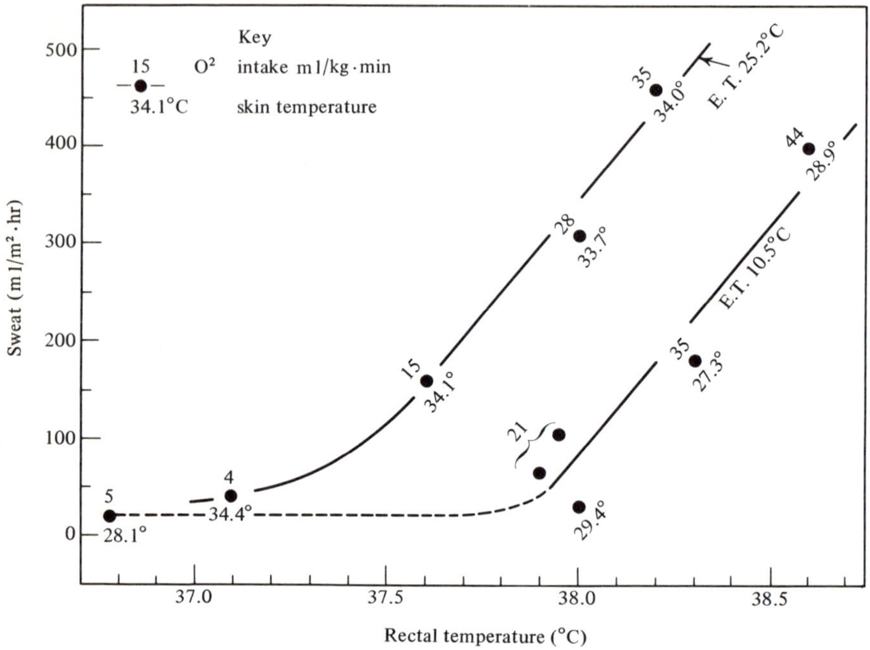

FIGURE 5-14. Relation of rate of sweating to increasing rectal temperature. The increments in rectal temperature were produced by increasing intensities of work as indicated by O_2 intake. Increasing work did not affect skin temperature. (From S. Robinson. Physiological adjustments to heat. In: *Physiology of Heat Regulation and the Science of Clothing*. Edited by L. H. Newburgh. Philadelphia: Saunders, 1949, Chap. 5.)

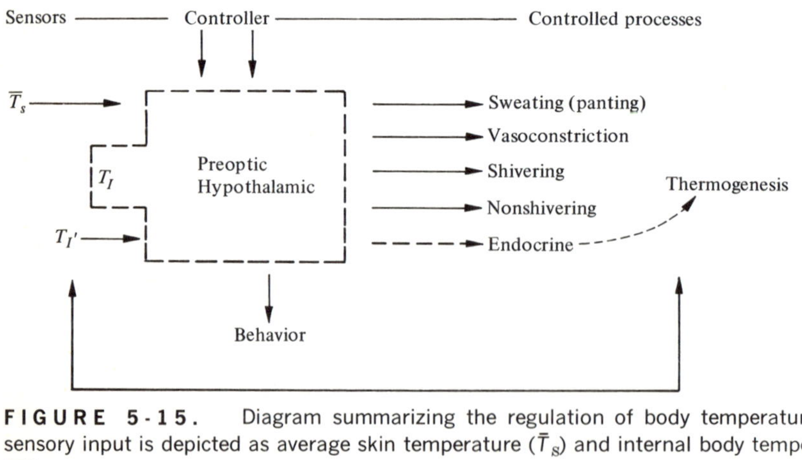

FIGURE 5-15. Diagram summarizing the regulation of body temperature. The sensory input is depicted as average skin temperature ($\bar{T}_S$) and internal body temperature (T_I).

or just rostral to it, there appear to be two main neural pathways to it. The first is from the skin receptors, and the second is the thermoceptive property of the hypothalamus per se. There is also evidence of a deep body thermoceptive area, perhaps in the spinal cord. Neurochemical effects of norepinephrine and 5-hydroxytryptamine have been described. The sensory input has been discussed. The thermoceptive property of the hypothalamus has been inferred from the relations of cranial temperature to the drive on controlled processes, such as heat production or sweating. The possibility of thermoceptive neurons is supported by single cell recordings in this area which give evidence for three classes of neurons—one class in which the firing rate is thermally insensitive, a second class in which the firing rate is inversely related to neurone temperature, and the third class in which firing rate is proportional to hypothalamic temperature (16). The hypothalamic area has many properties of a thermostat and it is tempting to characterize the deep body temperature as its "set point." The intrinsic reference seems not to be fixed but to have a range determined by some extrinsic factors. The hypothalamus also operates functionally as a source of releasing factors for the trophic hormones for endocrines, such as the thyroid, and the source of the sympathetic drive for adrenal medullary secretion of epinephrine and norepinephrine.

The neural controlled processes fall into four major categories: (1) skin vascular resistance, (2) sweating or panting, (3) shivering and nonshivering thermogenesis, and (4) behavior. A parallel system of neurohumorally controlled endocrine glands influences metabolism.

Factors that alter response characteristics may not only influence the hypothalamic regulatory neurons but may exert their effect by convergence on the efferent pathways.

The efferent or output components of the system converge on processes which subserve other systems (vasoconstriction for blood pressure; voluntary action in muscle). Some, such as sweating and panting, are more clearly closely linked in temperature regulation. Each of the controlled processes may be modified by local effects.

Fever

Fever is defined as an elevation of the body temperature above normal. Because individual body temperatures distribute themselves about a mean and body temperature varies during the day, the definition of an elevation of temperature is quantitatively difficult. Fevers that occur as a result of infection may be intermittent or may continue over days. Body dehydration also leads to an elevated temperature characterized as fever. The chemical agents that give rise to a fever are classed as pyrogens.

The pathogenesis of fever is intimately related to the mechanisms that control and regulate normal body temperature. At the onset of a fever, the general physiologic reactions are those of heat conservation, intense vasoconstriction, and possibly shivering. The person may have a "chill." Studies in animals with pyrogens demonstrate that these chemical substances activate the heat-conservation mechanisms. Studies with endotoxin have shown that as little as 0.0005 $\mu g/kg$ of the purified endotoxin administered intravenously will produce a fever. The febrile response occurs after a latent period which may be as long as 90 min. During this time in animals there is a marked fall in the circulating polymorphonuclear leukocytes. Their disappearance is attributed to adherence to the endothelial lining of blood vessels. The latent period suggests that the endotoxin may be converted to some other product that is involved in directly producing the fever. There does appear in blood, following the administration of endotoxin, a protein different from the endotoxin that can induce fever without a latent period. The response is directly related to dose. When the polymorphonuclear leukocytes are eliminated by the effect of nitrogen mustard, rabbits do not develop fever after the usual doses of endotoxin. These same animals can react to the administration of an endogenous pyrogen from an extraneous source.

Injections of small amounts of exogenous pyrogen into the hypothalamus produces a fever, but with the same lag time as a systemic injection. However, extracts of the leukocytes, endogenous pyrogen, when injected in the hypothalamus produce a fever without a lag period. The fever with an infection then seems directly related to an effect on the hypothalamus by an endogenous pyrogen released from body cells. This effect is specific for a prescribed region in the hypothalamus. (See Figure 5-8.)

When the fever terminates, the physiologic responses are all characteristic of activation of heat-loss mechanisms. There is a peripheral vasodilation and intense sweating.

During fevers that are noncyclic, the body appears to regulate about a new central temperature. When afebrile and febrile subjects are tested with a known heat load, the response and relationship between the peripheral loss of heat and the increase in core temperature are similar. During the rise in fever, the response is diminished. This would lead one to characterize a fever in general terms as a resetting of the intrinsic reference point for body temperature in the hypothalamic region. It is postulated that the response to endogenous pyrogen is mediated by norepinephrine or 5-hydroxytryptamine or that it may act directly at the neuronal level.

Diseases That Affect Temperature Regulation

RAYNAUD'S DISEASE AND RAYNAUD'S PHENOMENON

Maurice Raynaud's description of local asphyxia and symmetric gangrene of the extremities included two types of responses: one in which the digits were subject to attacks of syncope and asphyxia brought on by cold and sometimes emotion, and one in which gangrene occurred in cold, cyanotic, painful digits. The first of these (attacks of cyanosis with or without pallor involving the fingers of both hands) is now called Raynaud's disease; the second condition is called Raynaud's phenomenon. It has been demonstrated that persons with Raynaud's disease have reduced digital blood flows and that blood flow is more markedly reduced by exposure to cold. There is not an excessive amount of catecholamine in the tissue or venous blood. It is concluded that Raynaud's disease is a reflection of hypersensitivity of the digital arteries to local cold aided by the normal reflex pathways of response to cold exposure.

COLD HEMOAGGLUTINATION

The phenomenon of cold hemoagglutination has been described in humans and animals. The agglutinin is a gamma globulin which has been characterized serologically. The characteristics of the disease are agglutination of red cells by serum having cold agglutinin between 0 and 5°C, but rarely above 25°C. The agglutination is reversible on warming. The occurrence of cold hemoagglutinins is given as varying between 15 and 95 per cent. The incidence is higher in subjects who have had frostbite but is not inducible.

BURGER'S DISEASE (THROMBOANGIITIS OBLITERANS)

Burger's disease is an episodic and segmental inflammatory lesion of arteries and veins resulting in the permanent obliteration of their lumina. The effects are due solely to the obstruction.

CARDIOVASCULAR DISEASE

Heart failure is characterized by a striking reduction in the response to a variety of endogenous and exogenous vasodilator stimuli (such as exercise and reactive hyperemia). Circulating catecholamines and sympathetic vasoconstrictor activity are not solely responsible for the elevation of the peripheral vascular resistance and the reduced response to vasodilation stimuli (30). There may be an alteration of mechanical properties of the vessels. Thus, muscle pain and cold extremities will be symptoms in this disease.

In primary systemic amyloidosis there is a correlation between symptoms of peripheral vascular disease, anatomic evidence of arterial and arteriolar infiltration by amyloid, and decreased dilator capacity of the forearm resistance vessels (29).

SEQUELAE OF FROSTBITE AND COLD INJURY

Although persons who have suffered from frostbite or local cold injury are sensitive to cold, it is most likely due to anatomic pathology rather than physiologic changes.

IDIOPATHIC MALIGNANT HYPERTHERMIA DURING ANESTHESIA

Although of low incidence, the occurrence of hyperthermia during anesthesia is a striking event and has a high mortality. In cases described, some have a muscle rigidity but others do not. The temperature rises rapidly (as much as 6 to 8°C/hr) and rigorous cooling methods must be used. The increase seems primarily to be an increase in heat production but the mechanism is unknown. It does not resemble a pyrogen-induced fever, although the appearance of endogenous lymphocytic pyrogen has not been disproved.

HEAT ILLNESSES

Four categories of heat illnesses are generally recognized.

1. Skin disorders, including sunburn and prickly heat.
2. Heat syncope.
3. Heat exhaustion and heat cramps.
4. Heat stroke.

The latter three categories are distinguished by the physiologic mechanisms involved. Heat syncope results from the extensive peripheral vasodilation which may combine with orthostatic hypotension to produce cerebral ischemia.

Heat exhaustion and heat cramps are disorders or derangements of water and electrolyte metabolism.

Heat stroke is a disorder of thermoregulation. There is an absence of sweating and hyperthermia (usually over 41°C). Disturbances of brain function, such as unconsciousness, coma or convulsions, may be present. The prompt lowering of body temperature is axiomatic in the treatment of heat stroke.

It is well to remember that drugs which cause vasodilation or lowering of blood pressure will predispose an individual to heat syncope and that all sedative and tranquilizing drugs suppress or interfere with temperature regulation.

References

1. Andersson, B. Cold defense reactions elicited by electrical stimulation within the septal area of the brain in goats. *Acta Physiol. Scand.* **41:** 90–100, 1956.
2. Andersson, B., R. Grant, and S. Larsson. Central control of heat loss mechanisms in the goat. *Acta Physiol. Scand.* **37:** 261–280, 1956.
3. Aschoff, J. Significance of circadian rhythms for space flight. In: *Bioastronautics*

and the Exploration of Space. Edited by T. C. Bedwell, Jr., and H. Strughold. Washington, D.C.: Defense Document Center, 1965.

4. Bazett, H. C. The regulation of body temperatures. In: *Physiology of Heat Regulation and the Science of Clothing*. Edited by L. H. Newburgh. Philadelphia: Saunders, 1949, pp. 109–192.
5. Benzinger, T. H., C. Kitzinger and A. W. Pratt. The human thermostat. In: *Temperature: Its Measurement and Control in Science and Industry*. Edited by J. D. Hardy. New York: Reinhold, 1963, Part 3, Chap. 56.
6. Birzis, L., and A. Hemingway. Descending brainstem connections controlling shivering in cat. *J. Neurophysiol.* **19:** 37–43, 1956.
7. Brown, G. M., and J. Page. Effect of chronic exposure to cold on temperature and blood flow of the hand. *J. Appl. Physiol.* **5:** 221–227, 1952.
8. Carlson, L. D., and A. C. L. Hsieh. Cold. In: *The Physiology of Human Survival*. Edited by O. G. Edholm and A. L. Bacharach. New York: Academic Press, 1965, Chap. 2.
9. Cooper, K. E. Temperature regulation and the hypothalamus. *Brit. Med. Bull.* **22:** 238–242, 1966.
10. Cooper, K. E., W. I. Cranston, and A. J. Honour. Observations on the site and mode of action of pyrogens in the rabbit brain. *J. Physiol.* **191:** 325–337, 1967.
11. Cooper, K. E., W. I. Cranston, and E. S. Suell. Temperature regulation during fever in man. *Clin. Sci.* **27:** 345–356, 1964.
12. Cunningham, D. J., J. A. Stolwijk, N. Murakami, and J. D. Hardy. Responses of neurons in the preoptic area to temperature, serotonin and epinephrine. *Amer. J. Physiol.* **213:** 1570–1581, 1967.
13. Ferris, Jr., B. G., R. E. Forster, II, E. L. Pillion, and W. R. Christensen. Control of peripheral blood flow: responses in the human hand when extremities are warmed. *Amer. J. Physiol.* **105:** 304–314, 1947.
14. Folkow, R., R. H. Fox, J. Krog, H. Odelram, and O. Thoreu. Studies on the reactions of cutaneous vessels to cold exposure. *Acta Physiol. Scand.* **58:** 342–354, 1963.
15. Green, H. D., and J. H. Kepchar. Control of peripheral resistance in major systemic vascular beds. *Physiol. Rev.* **39:** 617–686, 1959.
16. Hellon, R. F. Thermal stimulation of hypothalamic neurones in unanesthetized rabbits. *J. Physiol.* **193:** 381–395, 1967.
17. Hemingway, A., and D. G. Stuart. Shivering in man and animals. In: *Temperature: Its Measurement and Control in Science and Industry*. Edited by J. D. Hardy. New York: Reinhold, 1963, Part 3, Chap. 36.
18. Hensel, H. Physiologie der thermoreception. *Ergebn. Physiol.* **47:** 166–368, 1952.
19. Keller, A. D. The role of circulation in the physiology of heat regulation. *Phys. Ther. Rev.* **30:** 511–519, 1950.
20. Krog, J., B. Folkow, R. H. Fox, and K. L. Andersen. Hand circulation in the cold of Lapps and North Norwegian fishermen. *J. Appl. Physiol.* **15:** 654–658, 1960.
21. Lewis, T. Observations upon the reactions of the vessels of the human skin to cold. *Heart* **15:** 177–208, 1930.
22. Nakayama, T., H. T. Hammel, J. D. Hardy, and J. S. Eisenman. Thermal stimulation of electrical activity of single units of the preoptic region. *Amer. J. Physiol.* **204:** 1122–1126, 1963.

23. Robinson, S. Physiological adjustments to heat. In: *Physiology of Heat Regulation and the Science of Clothing.* Edited by L. H. Newburgh. Philadelphia: Saunders, 1949, pp. 193–231.
24. Robinson, S., E. S. Turrell, H. S. Belding, and S. M. Horvath. Rapid acclimatization to work in hot climates. *Amer. J. Physiol.* **140:** 168–176, 1943.
25. Simon, E., W. Rautenberg, and C. Jessen. Initiation of shivering in unanesthetized dogs by local cooling within the vertebral canal. *Experientia* **27:** 447–480, 1965.
26. Simon, E., W. Rautenberg, R. Thauer, and M. Iriki. Die auslösung von kältezittern durch lokale kühlung im wirbelkanal. *Pflüger. Arch. Physol.* **281:** 309–331, 1964.
27. Speelman, C. R. Effect of ambient air temperature and of hand temperature on blood flow in hands. Am. J. Physiol. **145:** 218–222, 1945.
28. Wit, A., and S. C. Wang. Temperature sensitive neurons in preoptic/anterior hypothalamic regions: Effects of increasing ambient temperature. *Amer. J. Physiol.* **215:** 1151–1159, 1968.
29. Zelis, R., D. T. Mason, and W. Barth. Abnormal peripheral vascular dynamics in systemic amyloidosis. *Ann. Intern. Med.* **70:** 1167–1172, 1969.
30. Zelis, R., D. T. Mason, and E. Braunwald. A comparison of vasodilator stimuli on peripheral resistance vessels in normal subjects and in patients with congestive heart failure. *J. Clin. Invest.* **47:** 960–970, 1968.

Additional Reading

DuBois, E. F. *Fever and the Regulation of Body Temperature.* Springfield: Thomas, 1948.

Hardy, J. D., ed. *Temperature: Its Measurement and Control in Science and Industry,* Part 3, New York: Reinhold, 1963.

Newburgh, L. H., ed. *Physiology of Heat Regulation and the Science of Clothing.* Philadelphia: Saunders, 1949.

Martin, P., R. B. Lynn, J. G. Dible, and J. Aird. *Peripheral Vascular Disorders.* London: Livingstone, 1956, Chap. 16.

Selle, W. A. *Body Temperature: Its Changes with Environment, Disease and Therapy.* Springfield: Thomas, 1952.

Shepherd, J. T. *Physiology of the Circulation in Human Limbs in Health and Disease.* Philadelphia: Saunders, 1963, Chap. 23.

Leithead, C. S., and A. R. Lind. *Heat Stress and Heat Disorders.* Philadelphia: Davis, 1964.

CHAPTER

6

Adaptation

Stress and Strain

In environmental physiology it is convenient to use the word *stress* to mean any change in the external environment that leads to a change in the internal environment of the unadapted animal. Strain in the animal results not only from the change in its internal environment but also from the responses of the regulatory systems attempting to maintain the *status quo ante*. As changes in the unadapted animal are used as reference points, reduction in strain will be a measure of the degree of adaptation.

There is an inverse relationship between the degree of strain that an animal can withstand and the length of time it is exposed to stress. This observation has led to the development of time-related indexes for adaptation (for example, time for body temperature or pulse rate to reach certain levels, time to exhaustion, time to fainting, time to death, and so on).

The control systems for regulation of temperature and of energy exchange incorporate both physical and chemical response mechanisms. These mechanisms respond in time to acute changes in energy requirement. If the disturbance to the system which gives rise to the error signal persists, the response systems may change in character. This change over a period of time is referred to as adaptation, acclimatization, or acclimation.* A special case of adaptation

*Hart suggests the following definition for these three terms: *adaptation*—alteration over many generations, evolutionary; *acclimatization*—changes induced by climate; *acclimation*—changes induced by temperature.

which can be specifically attributed to a diminution of a response to a fixed stimulus (in neurophysiologic terms) is called habituation.

Adaptation to Heat

CHANGES IN BODY TEMPERATURE, HEART RATE, AND SWEAT RATE

Adaptation to heat involves the series of physiologic adjustments that occur when an animal is transferred from a normal or cool environment to a hot environment. The most convenient way to describe the strain is in terms of a physiologic response to a fixed environmental condition. Rectal temperature, skin temperature (and, consequently, change in body heat content), and heart rate are the responses commonly used (8). The change in these variables with time indicate that there is a reduction in strain on the system after repeated exposures to heat stress (Figure 6-1).

When rectal temperature, heart rate, and sweat rate are plotted against exposure time, as in Figure 6-2, it is found that the reduction in strain can be accounted for by the increase in sweat rate. In the experiments depicted in the figure, heat storage of the men at day 9 is nearly 80 kcal less than at day 1. The difference in sweat rate accounts for the entire difference. The experiments also point to intriguing questions concerning changes in the regulatory system. In general terms, for any given skin or central temperature there is a greater sweat rate with succeeding exposures up to about the ninth day.

Acclimatization to heat can be accelerated and enhanced if the subjects exercise in the heat. Acclimatization may also be brought about by raising the rectal temperature.

Although a number of other physiologic systems have been implicated in the acclimatization process, none have been shown to be so fundamental as the adjustment of mean skin temperature to a level which permits thermal equilibrium between the body and the environment, maintaining the internal gradient to permit the transport of heat to the surface without overtaxing the circulation. These conditions generally are attained almost wholly as a result of increased evaporative cooling from increased sweat production.

CHANGES IN OTHER REGULATORY SYSTEMS

It is tempting to involve the endocrine system in the heat acclimatization process because of the marked load placed on both body fluid and electrolyte balance. However, none of the endocrine controls of fluid or electrolyte have been shown to be essential.

There have also been reports of increased blood volume with exposure

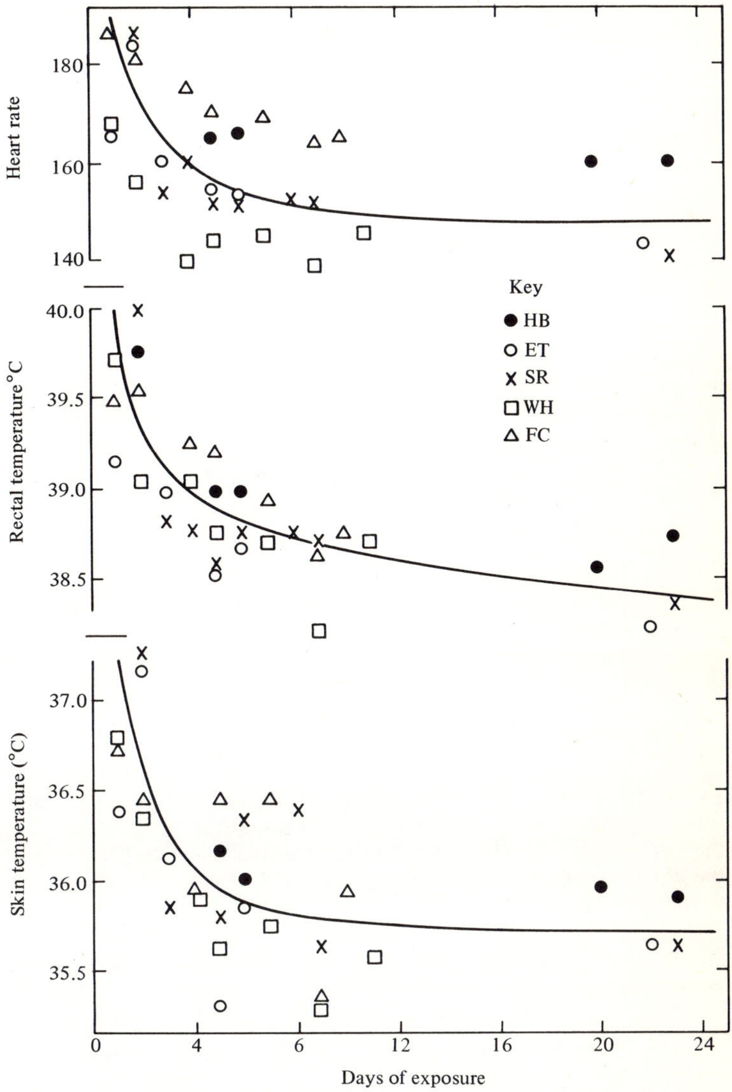

FIGURE 6-1. Acclimatization to heat resulting from daily exposure to a room maintained at 40°C with 23 per cent relative humidity. Final heart rate and rectal temperatures are reduced appreciably as a result of fall in the skin temperature at which adequate sweat is produced. (From S. Robinson, E. S. Turrell, H. S. Belding, and S. M. Horvath. Rapid acclimatization to work in hot climates. *Amer. J. Physiol.* **140:** 168–176, 1943.)

to heat. This would augment blood flow to the skin. But it has not been shown that circulation to the skin is the limiting factor in heat acclimatization. As mentioned, the primary process appears to be maintenance of skin temperature at a low enough level to provide for heat transport.

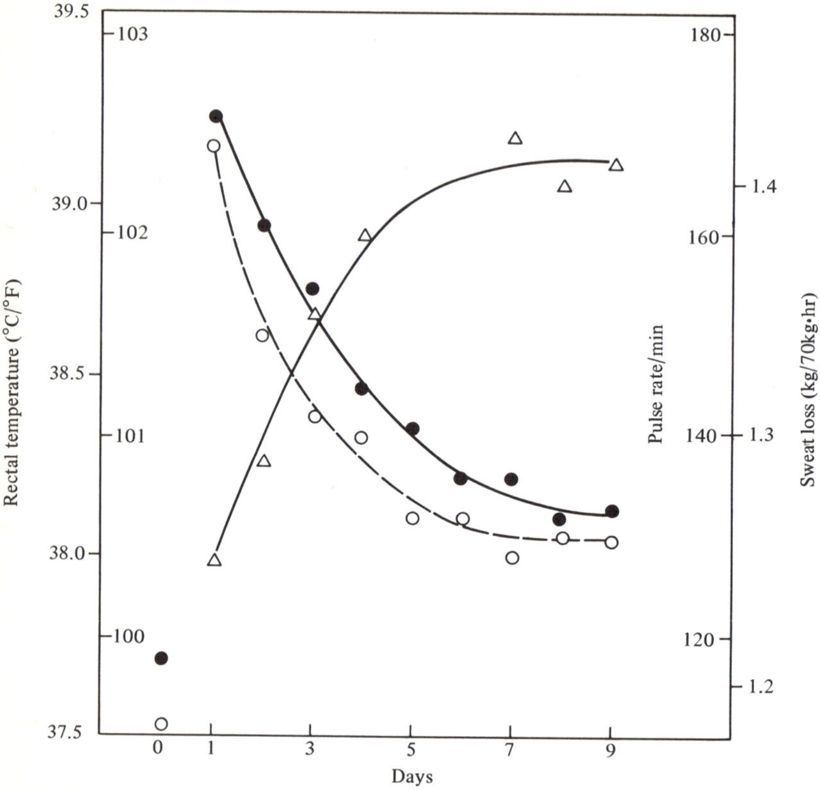

FIGURE 6-2. Typical average rectal temperatures (●), pulse rates (○), and sweat rates (△) of a group of men during the development of acclimation to heat. On day 0, the men worked for 100 minutes at an energy expenditure of 300 kcal/hr in a cool climate; the exposure was repeated on days 1 to 9 but in a hot climate with dry- and wet-bulb temperatures of 48.9 and 26.7°C.

LIMITS OF THE REGULATORY SYSTEM

The limits of the system are usually expressed as tolerance time. Values related to temperature, humidity, air movement, and activity are given in Figure 4-16 (page 67). The physiologic factors that determine these limits are complex, but one must be the maximum sweat rate and another the extent to which elevations in body temperature can be tolerated. Tolerance of heat certainly changes with disease, particularly those diseases which affect the circulatory system or water-balance mechanisms.

Adaptation to Cold

The signs and symptoms of adaptation to cold have been sought by looking for deviations in mean values of variables whose normal distribution has been established in persons not previously exposed to cold—for example, BMR, rectal temperature, food intake, skin and body temperature, weight–surface area relationship, body composition, manual dexterity. Secondly, indexes of adaptation have been sought in changes in the patterns of physiologic responses to cold stress—for example, patterns of heat loss, blood flow, heat production, and endocrine activity.

Man lives in many climates and there are mirked differences in the social and cultural adjustments to his environment. There have been a number of efforts to determine physiologic differences in response to cold among various ethnic groups and to ascertain that these differences are genetic. The greatest problem in interpreting the physiologic and morphologic adjustments to environmental extremes is to decide whether they are plasticity responses or the result of selective processes. There appears to be no conclusive evidence of an ethnic or a racial difference in response to cold exposure.

CHANGES IN BASAL METABOLIC RATE

Basal metabolic rate is negatively correlated with average temperature of the climate. The significance of this is not entirely ethnic, because the correlation exists within ethnic groups over wide climatic range and a seasonal effect has also been shown. Body weight is also negatively correlated with the average temperature of the climate; paradoxically, skinfolds indicate less subdermal fat in the normal groups, indicating that chronic exposure to cold leads to an increase in the ratio of active tissue mass to total body weight.

When heat production is plotted against skin temperature, as in Figure 6-3, there are indications of a reduced responsiveness of the heat-production system to skin cooling after acclimatization to cold. In contrast to this method of study, experiments have been conducted on overnight cooling and heat production. The regression lines of heat production on average skin temperature from a number of these studies are shown in Figure 6-4. It is interesting to note that the Alacaluf Indian and the Australian Aborigine, who were able to sleep during the overnight cooling, did not respond to cooling of the skin from about 32°C to 27.5°C by an increase in heat production. Although the two groups are similar in their lack of response to cooling of the skin, the Alacaluf Indian has a metabolic rate about 40 per cent higher than that of the Aborigine.

CHANGES IN PERIPHERAL BLOOD FLOW

Many studies have been made of people whose hands are chronically exposed to cold. In general, the deduction can be made that blood flow and temperature in the hands of these individuals are maintained at higher levels than in controls. The Japanese and Korean Ama show striking seasonal changes in their response to cold.

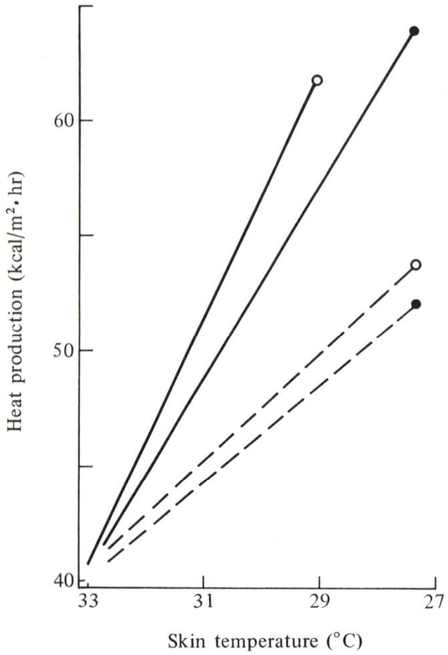

FIGURE 6-3. Open circles represent heat production after 1 hour cold exposure of clothed men, before (solid line) and after (broken line) 2 weeks of daily cold exposure. Closed circles represent heat production after 1 hour cold exposure of seminude men, in summer (solid line) and winter (broken line). (From L. D. Carlson and A. C. L. Hsieh. Cold. In: *Physiology of Human Survival.* Edited by O. G. Edholm and A. L. Bacharach. New York: Academic Press, 1965, Chap. 2.)

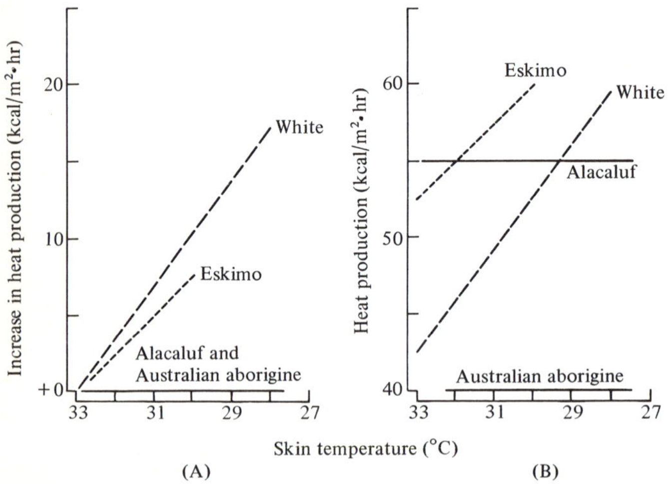

FIGURE 6-4. A: The increase in heat production and B: the total heat production of Caucasians, Eskimos, Alacaluf Indians, and Australian Aborigines plotted against skin temperature during overnight sleeping experiments. (From L. D. Carlson, and A. C. L. Hsieh. Cold. In: *Physiology of Human Survival.* Edited by O. G. Edholm and A. L. Bacharach. New York: Academic Press, 1965, Chap. 2.)

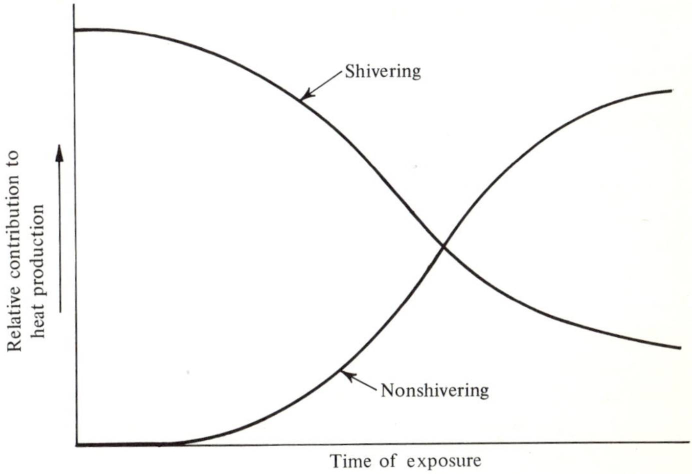

FIGURE 6-5. Schematic diagram of the changeover from shivering to nonshivering heat production with time of exposure.

CHANGES FROM SHIVERING TO NONSHIVERING HEAT PRODUCTION

The most striking shift in physiologic responses which have been noted with exposure to cold is the change from shivering to nonshivering thermogenesis. Figure 6-5 depicts schematically the changeover with time of exposure.

Shivering thermogenesis is mediated through pathways which activate skeletal muscle. Nonshivering thermogenesis seems to be mediated by the autonomic nervous system. Norepinephrine is calorigenic in cold-adapted animals. The actual target organ is not definitely known; brown adipose tissue is the most significantly affected.

The evidence for nonshivering thermogenesis comes in part from Keller's observations on gross shivering. He emphasized that cold-stimulated nonshivering heat production in dogs is just as distinct an entity as shivering heat production. Each has a separate and distinct group of nerve cells and a unique descending fiber tract in the hypothalamic gray matter and brain stem. Each is equally dependent on impulses from its own efferent neurons. Electromyographic evidence also establishes the existence of nonshivering thermogenesis as shivering decreases with continued exposure to cold.

Hardy and DuBois (3) found a true chemical regulation* in some women studied and stated that the lack of significant change in the metabolism of women and some men with a change in the environment shows that there must be a chemical regulation which increases metabolism in the core of the

*Because of confusion in the use of this term since Rubner defined it, *nonshivering thermogenesis* is a better term.

body to compensate for the lowering of metabolism in the periphery.

Nonshivering thermogenesis was first clearly demonstrated in the rat by the use of curare to block shivering and was found to be mediated by the sympathetic nervous system by the use of blocking agents. It appears to exist in varying degrees in the rat, rabbit, dog, and man but not in the bird or the miniature pig, and cold acclimation enhances this avenue of heat production. This change requires days or weeks and is definitely present in the rat after three weeks' exposure to 5°C.

One of the unanswered questions in this field of research is how the same stimulus, cold, can result in shivering when the animal is first exposed but in nonshivering thermogenesis after a few weeks of exposure. In shivering, the heat is produced near the surface and there is an increase in convective heat loss resulting from concomitant skin movement. There is, thus, a teleologic reason for the development of nonshivering thermogenesis, which is less wasteful of heat. Attempts have been made to explain the shift on the basis of a shift in metabolic pathways. Because the neural control of the two systems is different, the shift would have to be initiated in the central nervous system.

Newborn animals (with the exception of the bird and the miniature pig) have characteristic nonshivering thermogenesis, which is gradually displaced by shivering thermogenesis as the animal grows. Figure 6-6 is a schematic representation of this event. The time course of nonshivering thermogenesis is of interest, then, during the life span of the animal and also during the time of exposure to a cold environment. In the newborn the increase in oxygen consumption with cold exposure is predominantly nonshivering (norepinephrine mediated as demonstrated by sympathetic blocking agents), but as the

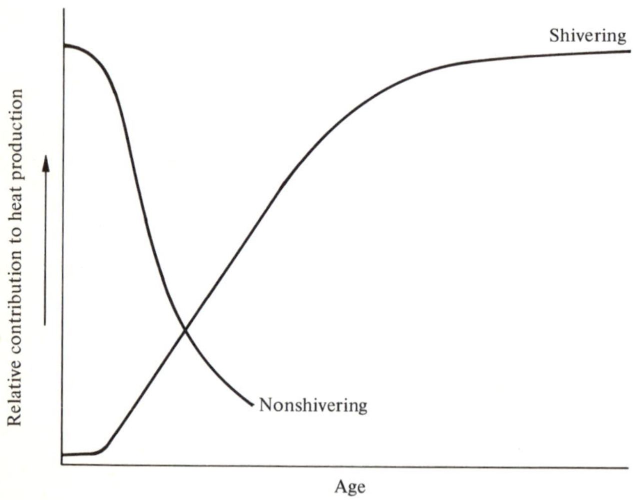

FIGURE 6-6. Schematic diagram of the reduction of nonshivering thermogenesis and the increase in importance of shivering as the animal grows older.

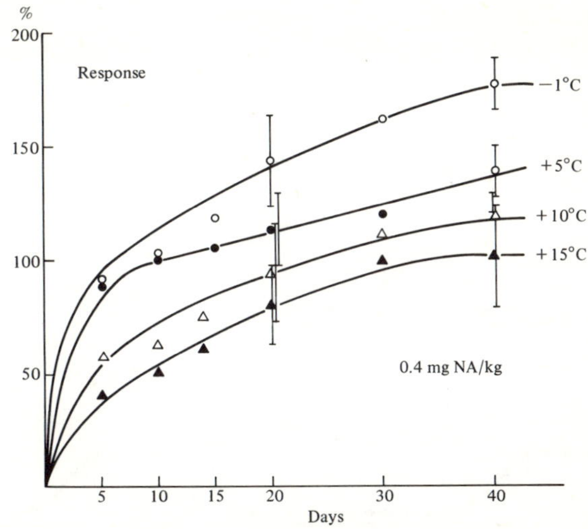

(A) The nonshivering thermogensis (as measured by the metabolic response to infusion of norepinephrine) develops in time and is proportional to the degree of cold exposure.

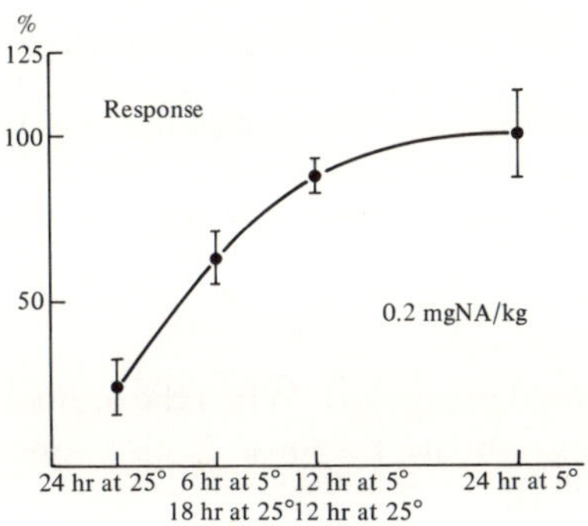

(B) The nonshivering thermogenesis developed is proportional to the amount of cold exposure in each day.

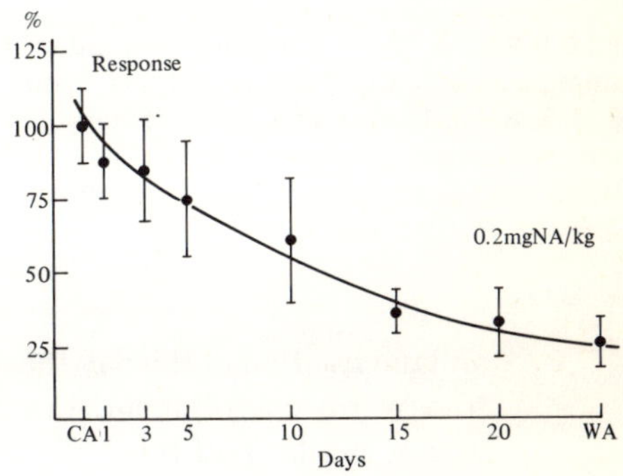

(C) The nonshivering thermogenesis decays in time following cold exposure.

FIGURE 6-7. The maximum metabolic responses of rats after administration of norepinephrine. (From L. Jansky, R. Bartunkova, and E. Zeisberger. Acclimation of the white rat to cold: Noradrenaline thermogenesis. *Physiol. Bohemoslov.* **16:** 366–372, 1967.)

animal grows, the cold-induced oxygen consumption from nonshivering thermogenesis decreases to zero unless the animal is continuously exposed to a cold environment. This is an interesting teleologic function in that the animal is prepared for a cold environment. The nonshivering thermogenesis is replaced by shivering thermogenesis. The temporal course of these effects is related to the amount of brown fat in the guinea pig. In the adult animal the nonshivering thermogenesis can be returned to function by cold exposure. The extent to which norepinephrine induces thermogenesis is dependent on the duration of cold exposure (half-time of about 8 to 10 days), the coldness,

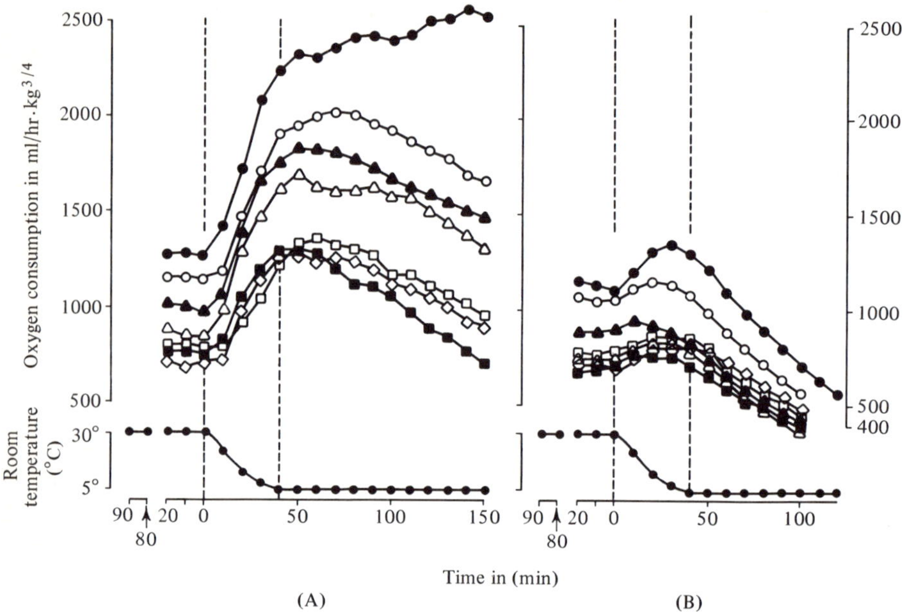

FIGURE 6-8. Oxygen consumption of curarized rats with room temperature cooled from 30 to 5°C. (*A*) Cold-adapted rats, symbols from top down indicate control, and then 2, 4, 6, 8, 10, 12 days after thyroidectomy. (*B*) Warm-adapted rats, symbols from top down indicate control, and then 4, 8, 12, 16, 20, 24 days after thyroidectomy. (From A. C. L. Hsieh, and L. D. Carlson. Role of the thyroid in metabolic response to low temperature. *Amer. J. Physiol.* **188:** 40–44, 1957.)

and the fraction of the day the animal is exposed. When the animal is returned to a warm environment, the decay in the response is also time dependent with a similar half-time (Figure 6-7).

CHANGES IN THE THYROID GLAND

It has been shown that cold acclimation is associated with an increase in thyroxin turnover and thyroxin requirement. The increase in turnover has been illustrated by Cottle and Carlson (2). The increased requirement has been shown by substitution techniques by Woods and Carlson (10). Ingbar and Bass (5) have noted increases in thyroxin utilization in nude men exposed to 14° C for two weeks. Rats in the cold can survive with lesser amounts of thyroxin, but Hsieh has shown that they do much better when given supplemental thyroxin.

The calorigenic action of epinephrine is synergistic with thyroxin. (See Figure 2-15, page 29). But blood thyroxin does not appear to be increased. Further elucidation of the role of thyroxin is derived from the work of Hsieh and Carlson depicted in Figure 6–8. Cold-induced nonshivering thermogenesis progressively decreased with time following thyroidectomy, but the response persisted. It appears that in the absence of thyroid hormone the animal cannot achieve the metabolic rates necessary to maintain thermal equilibrium in the cold (4).

CHANGES IN THE ADRENAL CORTEX

By tests, such as of the ability to survive at very low temperatures, adrenocortical hormones do not appear to be essential to the maintenance of cold acclimation, although they do appear to be required in greater than normal amounts on initial exposure to cold. Once cold acclimation is established, it can be maintained by a constant and relatively low rate of administration of adrenocortical hormones. The actual production or turnover of these hormones in the cold-acclimated animal is not known.

References

1. Carlson, L. D. Nonshivering thermogenesis and its endocrine control. *Fed. Proc.* **19:** 25–30, 1960.
2. Cottle, M., and L. D. Carlson. Turnover of thyroid hormone in cold-exposed rats determined by radioactive iodine studies. *Endocrinology* **59:** 1-11, 1956.
3. Hardy, J. D., and E. F. DuBois. Differences between men and women in their responses to heat and cold. *Proc. Nat. Acad. Sci.* **26:** 389–398, 1940.
4. Hsieh, A. C. L., and L. D. Carlson. Role of the thyroid in metabolic response to low temperature. *Amer. J. Physiol.* **188:** 40–44, 1957.
5. Ingbar, S. H., and D. E. Bass. The effects of prolonged exposure to cold on production and degradation of thyroid hormone in man. *J. Endocrinol.* **15:** ii–iii, 1957.
6. Jansky, L., R. Bartunkova, and E. Zeisberger. Acclimation of the white rat to cold: Noradrenaline thermogenesis. *Physiol. Bohemoslov.* **16:** 366–372, 1967.
7. Keller, A. D. Hypothermia in the unanesthetized poikilothermic dog. In: *The Physiology of Induced Hypothermia.* Edited by R. D. Dripps. Washington, D. C.: National Academy of Sciences, Nat. Res. Coun. Publication 451, 1956, pp. 61–79.
8. Robinson, S., E. S. Turrell, H. S. Belding, and S. M. Horvath. Rapid acclimatization to work in hot climates. *Amer. J. Physiol.* **140:** 168–176, 1943.
9. Sellers, E. A., and S. S. You. Role of the thyroid in metabolic responses to a cold environment. *Amer. J. Physiol.* **163:** 81–91, 1950.
10. Woods, R., and L. D. Carlson. Thyroxine secretion in rats exposed to cold. *Endocrinology* **59:** 323–330, 1956.

Additional Reading

Carlson, L. D. Reactions of man to cold. In: *Medical Climatology.* Edited by S. Licht. New Haven: Licht, 1964, pp. 196–228.

Carlson, L. D., and A. C. L. Hsieh. Cold. In: *The Physiology of Human Survival.*

Edited by O. G. Edholm and A. L. Bacharach. New York: Academic Press, 1965, pp. 15–51.

Fox, R. H. Heat. In: *The Physiology of Human Survival.* Edited by O. G. Edholm and A. L. Bacharach. New York: Academic Press, 1965, pp. 53–79.

Hardy, J. D., ed. Temperature. Its measurement and control in science and industry. In: *Biology and Medicine.* London: Reinhold, 1963, vol. III, part 3.

Leithead, C. S., and A. R. Lind. *Heat Stress and Heat Disorders.* Philadelphia: Davis, 1964.

Lind, A. R. Physiologic responses to heat. In: *Medical Climatology.* Edited by S. Licht. New Haven: Licht, 1964, pp. 164–195.

APPENDIX

I

A Standard System of Symbols and Units for Thermal Physiology*

THERE IS A CONTINUOUS EFFORT by all scientific societies to standardize definitions and symbols used in their publications. The tables in this section are based on proposals made by the Subcommittee on Thermal Physiology of the International Union of Physiological Sciences, Dr. J. D. Hardy, chairman. The units adopted are based on the recommendations of the 11th General Conference on Weights and Measures in 1960. The International System of Units (SI) has six basic units: meter (m), kilogram (kg), second (s), ampere (A), degrees Kelvin (°K), and candela (cd). Table A-1 gives supplementary and derived units which are of particular interest to thermal physiologists.

*The tables in Appendix I are modified from *J. Appl. Physiol.* **27**: 439–445, 1969.

TABLE A-1
Supplementary Basic Units for Use in Thermal Physiology

Name	Unit*	Abbreviation of Unit
Plane angle	radian	rad
Solid angle	steradin	sr
Area	square meter	m^2
Volume†	cubic meter	m^3
Frequency	hertz	Hz (s^{-1})
Density	kilogram per cubic meter	kg/m^3
Velocity	meter per second	m/s
Force	newton	N ($kg \cdot m/s^2$)
Pressure‡	newton per square meter	N/m^2
Work, energy, quantity of heat	joule	J ($N \cdot m$)
Power	watt	W (J/s)

*Units do not take a plural form.
†The liter (l) is a unit approved for use in conjunction with the SI.
‡mm of Hg or torr will continue as an acceptable unit for pressure with the SI until steam tables and other related vapor pressure tables are available more generally in N/m^2.

TABLE A-2
Principal Physical Quantities

Quantity	Symbol*	Unit	Abbreviation of Unit
Absorptance (radiation) (5)**	α	N. D.	
Area	A	square meter	m^2
Conductivity, thermal (3)	k	watt per meter per degree Celsius	W/(°C·m)
Density	ρ	kilogram per cubic meter	kg/m^3
Emittance (radiation)	ϵ	N. D.	
Heat, rate of exchange	H	watt per square meter	W/m^2
Heat, latent	λ	joule per kilogram	J/kg
Heat, quantity (energy)	J	joule	J
Heat-transfer coefficient (total sufrace conductance) (6)	h	watt per square meter per degree Celsius	W/(m^2·°C)
Irradiance (incident radiant flux density)	I	watt per square meter	W/m^2
Length	L	meter	m
Mass	m	kilogram	kg
Mass transfer rate (2)	$\dot{m}$	kilogram per second	kg/s
Pressure (partial)	P	newton per square meter or millimeter of Hg	N/m^2
Reflectance (radiation)	ρ	N. D.	

Quantity	Symbol*	Unit	Abbreviation of Unit
Resistance, thermal (insulation)	R; I	degree Celsius and square meter per watt	°C·m²/W
Specific heat	c	joule per kilogram per degree Celsius	J/(kg·°C)
Time (8)	t	second	s
Temperature (9)	T	degree Kelvin	°K
		degree Celsius	°C
Transmittance (radiation)	τ	N. D.	
Ventilation rate (2)	$\dot{V}$	cubic meter per second; liter per second	m³/s, l/s
Velocity, linear	v	meter per second	m/s
Volume	V	cubic meter, liter, milliliter	m³, l, ml
Work rate (7)	*W*	watt, joule per second	W, J/s

*When symbols are used in mathematical equations, they are usually shown in italic type.

**Numbers in parentheses are indicative of the following:

1. A bar over any symbol denotes mean value of the quantity.
2. A dot over any symbol denotes the time rate of change: *d/dt.*
3. The suffix *-ity* (for example, conductivity, emissivity, and so on) denotes a fundamental—that is "intrinsic state or condition of" the quantity—and one which is not dependent on external dimensions or conditions.
4. The suffix *-ive* (for example, *conductive*) denotes "the nature of."
5. The suffix *-ance* (for example, *reflectance*) denotes a specific form, condition, or modification of the fundamental quantity.
6. The proposal for writing units as indicated may be used with ease on a typewriter. The rigorous method—that is, W m^{-2}°C^{-1}— is acceptable but awkward for a typewriter; W/(m^2·°C) is also accurate and less difficult for a typewriter; W/m^2/°C is incorrect.
7. The kilopond meter per min is used frequently by Europeans for rate of work (*W*) on a bicycle ergometer: 100 kpm/min = 16.35 watt. When *W* has a value, all other quantities in the heat-balance equation must be expressed in the same units as *W*.
8. Whenever used, alternate abbreviations for time are suggested as *hr* for *hour,* and *min* for *minutes.*
9. Capital (or upper case) *T* is recommended for both Kelvin and Celsius. Confusion between the two meanings is unlikely. Degrees Kelvin must be used wherever *T* is raised to a power or whenever used in the gas equation ($pV = RT$). In all other cases, °C may be used. A lower-case *t* for deg Celsius leads to confusion with *t* for time.

TABLE A-3
Universal Constants*

Name	Symbol	Value
Avogadro constant	N_A	6.0225×10^{23} mol^{-1}
Boltzmann constant	k	1.38×10^{-23} J/°K
Gas constant	R	8.3143 J°/(K·mol)
Gravitational constant	G	6.670×10^{-11} N·m²/kg²
Stefan-Boltzmann constant	σ	5.67×10^{-8} W/(m²·°K⁴)

*Universal constants are always written in italics.

TABLE A-4
Physical Subscripts

Significance	Symbol	Example
ambient	a	T_a = ambient or dry-bulb temperature
conductive*	k	h_k = conductive-heat-transfer coefficient
convective*	c	h_c = convective-heat-transfer coefficient
diffusion	D	h_D = mass-heat-transfer coefficient by diffusion
emitting source	i	T_i = temperature of emitting source
evaporative	e	h_e = evaporative-heat-loss coefficient
pressure (constant)	P	c_P = specific heat at constant pressure
projected	p	A_p = projected area
radiation (radiative)	r	h_r = linear radiation-transfer coefficient
receiving surface	j	A_j = irradiated area
spectral wave length	λ	$F\lambda$ = spectral radiant flux intensity in band, $\lambda + \Delta\lambda$
volume (constant)	V	c_V = specific heat at constant volume
water vapor	w	P_w = partial pressure of water vapor
wet bulb	wb	T_{wb} = wet-bulb temperature

*See comment 4 to Table A-2 (page 109).

TABLE A-5
Physiologic Subscripts*

Significance	Symbol	Examples Example
arterial	ar	T_{ar} = arterial temperature
body	b	T_b = mean body temperature
blood	bl	V_{bl} = blood flow
esophageal	es	T_{es} = temperature of esophageal
expired	ex	T_{ex} = temperature of expired air
hypothalamic	hy	T_{hy} = hypothalamic temperature
inspired	in	T_{in} = temperature of inspired air
muscle	m	T_m = muscle temperature
oral	or	T_{or} = oral temperature
rectal	re	T_{re} = rectal temperature
skin	s	T_s = skin temperature
tympanic	ty	T_{ty} = tympanic temperature
venous	ve	T_{ve} = venous temperature

*In general, physiologic subscripts should have a lettered similarity to the property being described. The list of physiologic subscripts suggested is of necessity incomplete because each author will have a specific need to describe his particular heat-transfer problem. For example, an author may use T_{po} for preoptic temperature and T_{hp} for posterior hypothalamic temperature. Lower-case, two-lettered subscripts are encouraged for such descriptions.

TABLE A-6
Special Quantities Useful for Describing Heat Exchange

Quantity	Symbol	Unit
Ambient air temperature	T_a	°C
Mean radiant temperature (MRT)*	T_r	°K or °C
Operative temperature	T_o	°C
Effective radiant field (or flux) (+ for body warming)	H_r	W/m^2
Dimensionless factor (for example, shape factor)	F	always use with identifying subscript
Combined-heat-transfer coefficient	h	$W/(m^2 \cdot °C)$
Linear radiation-heat-transfer coefficient	h_r	$W/(m^2 \cdot °C)$
Convective-heat-transfer coefficient	h_c	$W/(m^2 \cdot °C)$
Insulation from the skin or clothing surface to a uniform environment	$T_a\ I_a$	$(°C \cdot m^2)/W$ in Clo units $= 0.155\ °C \cdot m^2/W$
Insulation of clothing	I_{cl}	
Relative humidity of ambient air	ϕ_a	N.D.
Useful Quantities Which Cannot Be Used in the Heat Balance Equation Are:		
Effective temperature	T_{eff}	defined in terms of T_a and T_{wb} (or ϕ_a)
Globe temperature	T_g	temperature of a Vernon black 6-in. diameter globe

*See comment 1 to Table A-2 (page 109).

TABLE A-7
Special Symbols for Convection and Mass Transfer

Quantity	Symbol	Unit
Diameter	d	m
Diffusivity, thermal	d	m^2/s
Diffusivity, mass	D	m^2/s
Enthalpy, specific	h	kJ/kg
Gas constant, water vapor	R_w	$3.47\ m^3 \cdot mm\ Hg/(kg \cdot °K)$
Gravitational constant	g	m/s^2
Heat, latent (enthalpy of evaporation)	λ	J/kg
Heat, specific (constant pressure)	c_p	$J/(kg \cdot °C)$
Heat-transfer coefficient by evaporation	h_e	$W/(m^2 \cdot mm\ Hg)$
Mass-transfer coefficient	h_D	m/s
Temperature difference	ΔT	°C
Vapor-pressure difference	ΔP	N/m^2; mm Hg; or torr
Viscosity, absolute	μ	$kg/(s \cdot m)$
Viscosity, kinematic	υ	m^2/s
Volume-expansion coefficient (gases)	β	1/°K

TABLE A-8
Useful Dimensionless Groups

Nusselt number, Nu = $h_c\ d/k$, or $h_c\ L/k$
Reynolds number, Re = $\rho\ v\ d/\mu$, or $\rho\ v\ L/\mu$ (= $v\ d/\upsilon$, or = $v\ L/\upsilon$)
Prandtl number, Pr = $c_p\ \mu/k = \upsilon/\alpha$ (0.72 for air between 10° and 50° C)
Grashof number, Gr = $(g\beta/\upsilon^2)\ \Delta Td^3$, or $(g\beta/\upsilon^2)\ \Delta TL^3$
Lewis number, Le = $h_c/h_D\rho c_P$
Schmidt number, Sc = $\mu/\rho D = \upsilon/D$
Sherwood number, Sh = $h_D\ d/D$, or $h_D\ L/D$

TABLE A-9
Conversion Factors for Selected Units in Current Usage to SI Units

To convert from	Abbreviation	To	Abbrev.	Multiply by
		Energy		
British thermal unit	Btu	joule	J	1055.9
calorie	cal	joule	J	4.187
foot-pound	ft-lb	joule	J	1.3558
kilocalorie	kcal	kilojoule	kJ	4.187
		joule	J	4187.
		Energy/(Area Time)		
Btu per square foot and hour	Btu/(ft²·hr)	watt per square meter	W/m²	3.1525
kilocalorie per square meter and hour	kcal/(m²·hr)	watt per square meter	W/m²	1.163
		Force		
kilogram force	kgf	newton	N	9.807
kilopond force	kp	newton	N	9.807
pound force (avoirdupois)	lb	newton	N	4.4482
		Power		
Btu per minute	Btu/min	watt	W	17.572
calorie per second	c/sec	watt	W	4.187
horsepower	1 hp = 550 ft·lb/sec	watt	W	745.
kilocalorie per hour	kcal/hr	watt	W	1.162
kilopond meter per minute	kpm/min	watt	W	0.1634
		Pressure		
inch of mercury	in Hg	newton per square meter	N/m²	3386.4
inch of water or (39.2° F)	in H_2O	newton per square meter	N/m²	248.8

To convert from	Abbreviation	To	Abbrev.	Multiply by
millimeter of mercury (0° C)	mm Hg	newton per square meter	N/m^2	133.3
millimeter of water (4° C)	mm H_2O	newton per square meter	N/m^2	9.806
torr	tr	newton per square meter	N/m^2	133.3
		Speed		
foot per minute	ft/min	meter per second	m/s	0.00508
foot per second	ft/s	meter per second	m/s	0.3048
kilometer per hour	km/hr	meter per second	m/s	0.2778
mile per hour	mph	meter per second	m/s	0.447
		Transfer Coefficients		
kilocalorie per square meter, hour and °C	$kcal/(m^2 \cdot hr \cdot °C)$	watt per square meter and °C	$W/(m^2 \cdot °C)$	1.163
British thermal unit per square foot, hour and °F	$Btu/(ft^2 \cdot hr \cdot °F)$	watt per square meter and °C	$W/(m^2 \cdot °C)$	5.67

APPENDIX

II

Measurement of Temperature

TEMPERATURE indicates the thermal energy level of a body. Because thermal energy will flow from a warm object to a cooler one, temperature is the property that determines the rate and direction of the flow of thermal energy or heat. Any function of matter that varies continuously and consistently with temperature may be used to indicate temperature. Three such functions are used routinely to measure temperature: thermal expansion, as in mercury, alcohol, and bimetallic thermometers; the change in electrical resistance of a conductor; and the Seeback effect at the junction of two dissimilar metals where an EMF dependent on temperature is generated.

All temperature-measuring devices use a system of units, temperature scales, to characterize the thermal state. The two fixed temperatures most widely used for reference points are the boiling and freezing points of water. The common scale in the British Empire and the United States is the Fahrenheit scale. The temperature range between the two fixed points is divided into 180 equal parts, Fahrenheit degrees. The value of 32 is assigned to the freezing point and 212 to the boiling point of water. The Celsius scale is preferred for most scientific work and in European countries. The Celsius scale divides the range between the freezing point, arbitrarily called zero, and the boiling point into 100 equal parts, Celsius degrees.

These temperature scales are limited by the inherent properties of the materials involved. For the most part, these ranges suffice in the conditions met in describing the effects of exposure to cold. However, in discussions of energy loss by radiation and in considerations involving thermodynamics such as gas volume changes, the absolute (A) or Kelvin (K) scale must be used.

This is, again, an arbitrary scale by which the size of the degree was chosen equivalent to the degree Celsius in this scale:

$$°K = °C + 273$$

When the zero point on this scale corresponds to conditions of zero energy, it represents the lowest temperature attainable.

Measuring Devices

Several measuring devices may be used: liquid thermometers, thermoelectric thermometers, resistance thermometers, radiometers, or bolometers. In physiologic measurements, liquid thermometers are primarily used to calibrate the other types. Their use is contraindicated in physiologic experiments because they must be read *in situ*; because, to be accurate to 0.1°C., they must be large and hence are slow to respond and withdraw heat from the tissues; and because they are fragile. Liquid thermometers standardized by the Bureau of Standards may be obtained from scientific supply houses and are recommended to standardize routinely used items of equipment. In certain instances this type of thermometer has the advantage of being usable as an indicator of minimum or maximum temperatures (for example, the clinical thermometer).

THERMOCOUPLES

The use of thermoelectric thermometers (thermocouples) was initiated by Seebeck's discovery in 1882 that in a closed circuit a small electromotive force is set up between a pair of bimetallic junctions when each is at a different temperature. In practice, one set of bimetallic junctions is kept in an ice bath at 0°C. The thermal-electromotive force values are small, about 40 microvolts per degree Centigrade for copper constantan junctions in the range of 0° to 100°C. The principle of this method indicates the need for care in using dissimilar metals in the external circuit as each becomes a thermocouple.

The choice of the metals to be used is determined by the range of temperatures to be measured. In the range pertinent to the discussions in this monograph, copper constantan and iron constantan junctions are the metals of choice with reference junctions maintained at 0°C. The EMF reverses sign at 0°C. As the temperature increases, the thermoelectric force eventually reaches the "neutral" temperature where it reverses direction and again approaches zero. Thus, the curve for the full range of temperature is parabolic and is expressed by the empirical equation

$$E = BT + (CT^2/2)$$

where E = electromotive force,
T = temperature,
and B and C = constants calculated for each pair of metals.

At higher temperatures, the thermoelectric force crosses zero at the "inversion" temperature. The selection of a specific thermocouple depends on the characteristics of this curve as well as on the physical properties of metals.

The EMF from thermocouples may be measured in a number of ways. (The reader is referred to Burton (2) for additional information.) The galvanometric method gives direct potential readings either by a high sensitivity galvanometer or through high-gain, direct-current amplifiers. The potentiometric system uses a galvanometer as a null-point indicator and the EMF is balanced against a standard EMF. This can be done with high accuracy on a Leeds and Northrup Type K-2 potentiometer, or automatically recorded on a Leeds and Northrup or Brown Instrument recorder. Remote recording is a major advantage of this type of thermometer, but the characteristics of the recording system must be carefully analyzed.

A second major advantage afforded by thermocouples is the small size of the measuring device. This varies from the fine thermocouples used by Webb (5) in measuring temperatures in the nasal airways to thermocouples with added mass.

In measuring skin temperatures a critical problem is to obtain the surface temperature without undue pressure on the skin and without conducting heat rapidly to the surrounding air. Further, the placement of skin thermocouples must be considered in light of Lewis' findings that the proximity of veins affects skin temperature measurements (3). A comparison of various surface-temperature-measuring devices is given in Table A-10. For determining rectal temperature, numerous thermocouple arrangements have been designed. The reliability of the use of rectal temperature has been discussed by Mead and Bommarito (4).

Subdermal temperatures are often determined with needle thermocouples. A fine thermocouple is threaded through a hypodermic needle and fixed with insulating varnish. The use of this type of thermocouple has been discussed by Bazett and McGlone (1), who determined the conductive loss along the needle and indicated the errors involved.

THERMISTORS

Thermocouples allow the use of a thermal EMF to determine temperature. Another electrical property may be used in measuring temperature, because the specific resistance of certain metals varies with temperature.

TABLE A-10
Comparative Evaluation of Electrical Temperature-Measuring Methods*

	Thermocouples	Thermopiles	Resistance Thermometers
1. Ease of construction	Easy	Difficult	Fairly easy
2. Cost of construction	Low	Fairly high	Low
3. Sensitivity	Limited by thermoelectric power of metals	Low	Not limited by thermoelectric power
4. Instrumental lag	Can be made very low	Slow unless vacuum thermopiles	Greater than thermocouples
5. Galvanometer required	Moderately sensitive, fairly portable	Quite sensitive	Relatively insensitive, rugged and portable
6. Possibility of	Not practical	Not easy	Easy
7. Errors likely	Parasitic Emf's sometimes troublesome; change in resistance of leads with temperature	Rise of cold junction temperature	Heat in coils caused by current; leads can be automatically compensated
8. Calibration	Closely linear in working range	Not linear; calculation required	Closely linear
9. Ease of measurement	Cold junction temperature must be read	Room temperature must be read	No observations except direct scale reading
10. Breakage	Fairly common with fine junctions; thermometers break	Not common; thermometers break	Fairly common; no thermometers to break
11. Interference with skin temperature	Possible if not properly made	None, no contact	Possible if current too high
12. Range of application	Good	Not under clothing	Good
13. Size of area measured	Can be very small; hard to average over large areas	Area not less than 1 cm^2	Area not less than 0.5 cm^2; gives average of large areas
14. Other possible application	Differential thermocouples for special purposes	Not developed, but possible	Differential temperature measurement; thermal stromuhrs (thermistors)

*From A. C. Burton. Temperature of skin: Measurement and use as index of peripheral blood flow. In: *Methods in Medical Research.* Edited by V. R. Potter. Copyright © 1948 Yearbook Medical Publishers, Inc., Chicago, vol. 1, pp. 146–166.

$$R_t = R_0\ (1 + a_0 t)$$

where R_t = resistance at temperature t,
R_0 = resistance at temperature 0°C,
a_0 = temperature coefficient of the metal referred to 0°C.

Three types of resistance thermometers may be found in common use:

1. Noble-metal resistance thermometer.
2. Base-metal resistance thermometer.
3. Semiconductor resistance thermometer or thermistors.

The platinum-wire resistance thermometer is an example of the noble metal resistance thermometer. This is used for measurements of temperature with a high degree of accuracy. A standard platinum thermometer is sold by Leeds and Northrup.

Nickel wire is most commonly used in base-metal resistance thermometers. The commercially available Dermalor is an example of this type of instrument. This type of resistance thermometer has use in calorimeters and heat-exchange meters, as will be discussed later.

In recent years the development of resistance units made of oxides of manganese, nickel, cobalt, copper, and uranium has added a new temperature-measuring device commonly called a thermistor. Thirty or more variations in thermistor types are described in *Bell Telephone Laboratory Bulletin*, May 1, 1947.

RADIOMETERS

Temperatures may also be measured by devices that indicate the amount of energy radiated from a body. These devices make use of thermocouples (which may be connected in series as a thermopile) or thermal resistance wire embedded in a suitable black body to absorb radiation. The instrument is necessarily limited to measurement of surface temperatures. The Hardy dermal radiometer is a commonly used device of this type.

References

1. Bazett, H. C., and B. McGlone. Temperature gradients in the tissues in man. *Amer. J. Physiol.* **87:** 415–451, 1927.
2. Burton, A. C. Temperature of skin: Measurement and use as index of peripheral blood flow. In: *Methods in Medical Research.* Edited by V. R. Potter. Chicago: Yearbook Medical Publishers, 1948, vol. 1, pp. 146–166.

3. Lewis, T. Observations upon the reactions of the vessels of the human skin to cold. *Heart* **15:** 177–208, 1930.
4. Mead, J., and C. L. Bommarito. The reliability of rectal temperatures as an index of internal body temperature. EPS Rep. No. 141, Q. M. Clin. Res. Lab., Lawrence, Mass., 1948.
5. Webb, P. P. The measurement of air temperature. *Rev. Sci. Instrum.* **23:** 232–234, 1952.

APPENDIX III

Calorimetry

CALORIMETRY is commonly divided into direct and indirect methods. Because demonstration indicates that the direct and indirect methods give similar results in the resting state, the direct method is seldom used because of its complexity. However, it must be noted that the two methods give similar results only in the resting state.

DIRECT CALORIMETRY

Direct calorimetry is accomplished according to the following principles, as given by Benedict and Carpenter (1). An example is shown in Figure A-1. The measurements of heat output by man as made by this apparatus are based on the fact that the subject is enclosed in a heatproof chamber through which a current of cold water is constantly passing. The amount of water, the flow of which for the sake of accuracy is kept at a constant rate, is carefully weighed. The temperatures of the water entering and leaving the chamber are accurately recorded at frequent intervals. The walls of the chamber are held adiabatic, thus preventing a gain or loss of heat by arbitrarily heating or cooling the outer metal walls. The withdrawal of heat by the water current is so controlled by varying the temperature of the ingoing water that the amount of heat brought away from the calorimeter exactly equals the heat output by radiation and conduction by the subject, thus maintaining a constant temperature inside the chamber. The latent heat of the water vaporized is determined by measuring directly the water vapor in the ventilating air current.

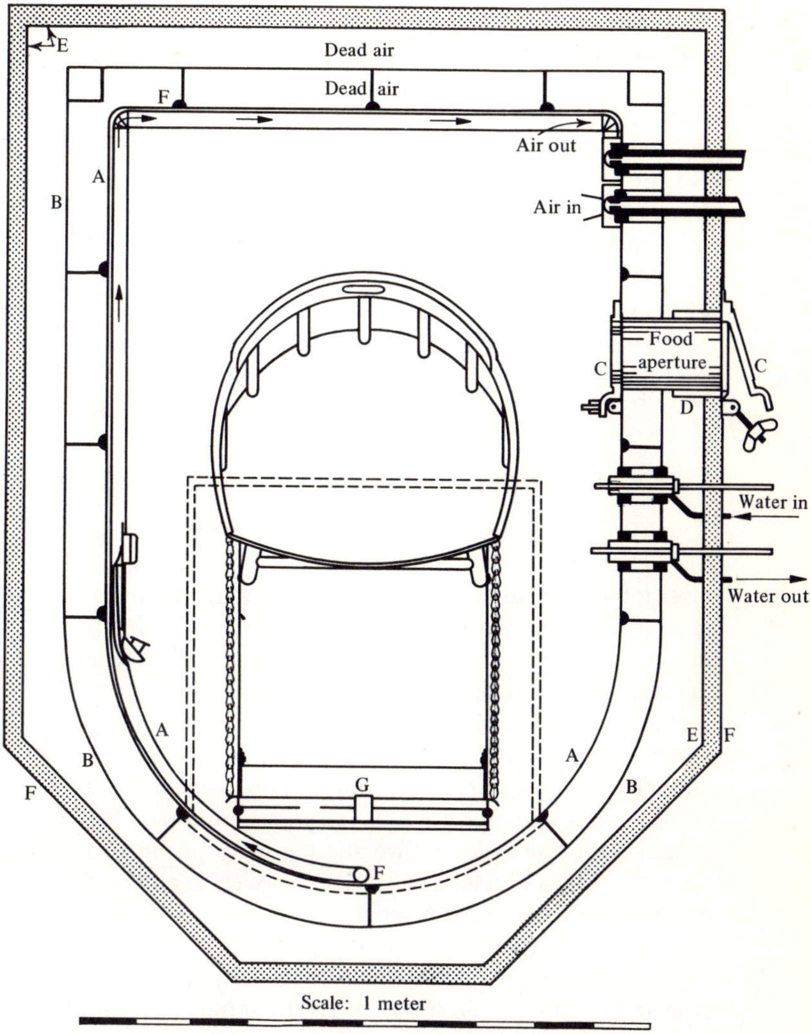

FIGURE A-1. Schematic diagram of the Atwater-Benedict calorimeter. (From F. G. Benedict and T. M. Carpenter. *Respiration Calorimeters for Studying the Respiration Exchange and Energy Transformations of Man.* Publ. No. 123. Washington, D.C.: Carnegie Institute of Washington.)

This conventional type of calorimeter can be combined with an indirect type. There are two serious drawbacks to the use of this type of calorimeter. In the first place, it serves as a heat trap, capturing the heat liberated by the subject and making no distinction among the physical means by which the heat is liberated. Secondly, the instrument as described is slow in response, and long-term experiments are necessary for adequate equilibrium and balance measurements.

Visser and Hodgson (11) and Carlson and coworkers (4) have described human calorimeters in which heat output is measured as a product of air flow rate and inlet to outlet air temperature difference. Figure A-2 is a schematic diagram of the calorimeter used by Carlson and coworkers. Air enters the floor and exits through the ceiling. Heat loss through the calorimeter walls is kept minimal by maintaining an outer room at calorimeter temperature. Heat transfer through the walls and door is measured with Hatfield Turner disks mounted on the walls.

Carrol and Visser (5) have devised a method for the direct measurement of the convective heat loss from a human body working in a natural posture under different environmental conditions in a climatic chamber. The method is based on the integration of the increase in heat in the wake behind the body by the use of temperature-integrating grids (Figure A-3).

PARTITION CALORIMETRY

Winslow, Herrington, and Gagge (14) designed an instrument (Figure A-4) that makes possible recording thermal interchanges as they occur within a brief interval of time, instead of summing up results extending over a period of several hours. Also, its data, given in terms of rate of heat change, are primary measurements in accordance with the physical avenues through which the interchange is occurring. They termed this new method of calorimetry *partition calorimetry*. The calorimeter was designed so that the various environmental conditions such as air temperature, air movement, atmospheric humidity, and mean radiant temperature could be controlled accurately and varied independently. For this purpose a nine-paneled booth of copper was constructed. The temperature and humidity of the air supplied to this booth and to the room in which it was located were regulated by the usual engineering procedures. The air temperature within the booth could be varied from 5 to 60°C, its relative humidity could be varied from 15 to 90 per cent. The air was introduced so that the velocity was minimal (15 to 25 ft/min). Higher velocities could be produced by using small fans located near the top of the booth. A new feature of the apparatus was the method of heating the chamber. Electrical elements located at the three openings in the chamber directed infrared radiant heat against the interior of the walls. Because copper reflects about 98 per cent of the infrared radiation, subjects seated in the booth received the full effect of hot radiation from the wall surfaces, while those surfaces themselves remained at approximately air temperature so that convection currents caused by hot walls were avoided. It was possible, for example, to obtain air temperatures of 5°C within the enclosure and to maintain a constant temperature of the walls close to that figure, while these same walls were reflecting radiant heat equivalent to that which would be given off by walls heated to 45°C. The use of this apparatus has contributed substantially to our knowledge of body heat loss.

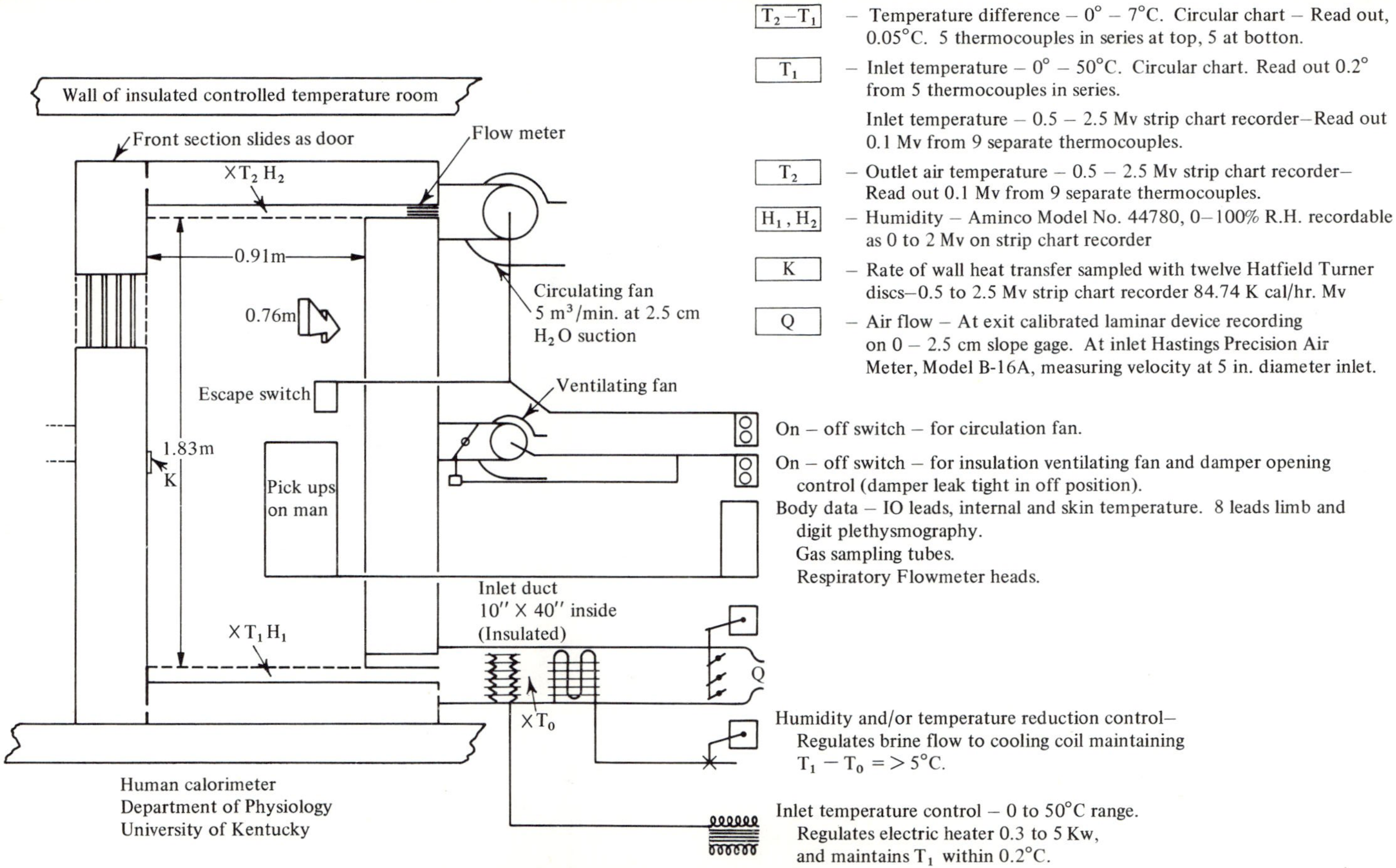

FIGURE A-2. Sagittal schematic of a calorimeter designed to indicate variations in body heat exchange and to allow change and stabilization of environment within a few minutes. (From L. D. Carlson, N. Honda, T. Sasaki, and W. V. Judy. A human calorimeter. *Proc. Soc. Exp. Biol. Med.* **117:** 327–331, 1964.)

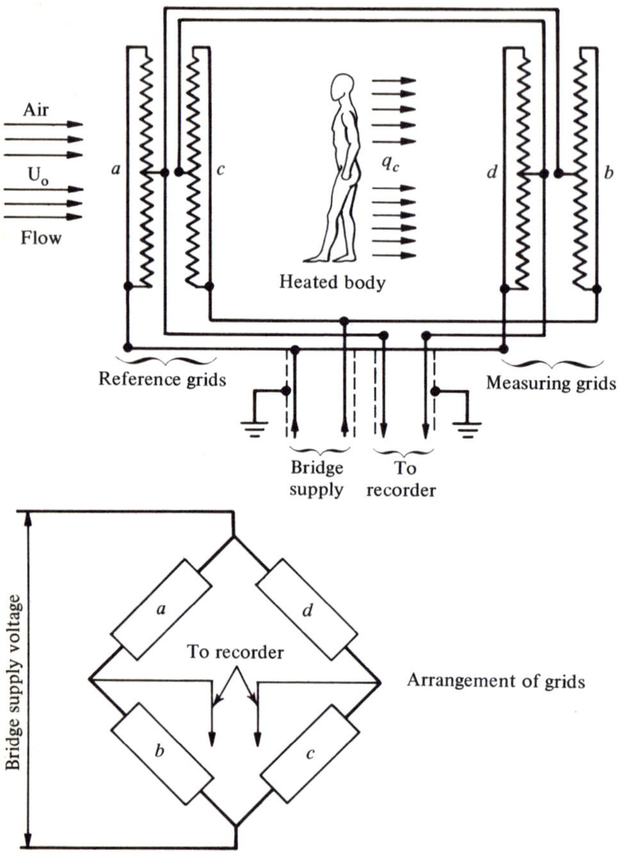

FIGURE A-3. Direct measurement of convective heat loss. (From D. P. Carroll and J. Visser. Direct measurement of convective heat loss from human subjects. *Rev. Sci. Instrum.* **37:** 1174–1180, 1966.)

THERMAL GRADIENT CALORIMETRY

Murlin and Burton (9), Day and Hardy (7), and Benzinger and Kitzinger (2) have designed calorimeters utilizing the thermal gradient principle, illustrated in Figure A-5. The method is based on the fact that when heat is transmitted across a layer of thermally conducting matter, a difference in temperature exists between the two surfaces of the layer. By an ingenious scheme of interlayering thermocouples above an insulating layer, a calorimeter was made which gave instantaneous measurements and recorded heat loss continuously. Such a calorimeter gives linear heat input and has a fast response time of about 42 sec.

This principle was also used by Young and coworkers (15) in designing an animal calorimeter as well as a thermal gradient suit for field test work. This calorimeter and thermal gradient suit, instead of using thermocouples across the thermal gradient layer, utilized thermal resistance wire set at specified intervals to measure the thermal gradient. The suit, for example, had nickel wire sewn on two sides of a standard layer of insulation. It had the advantage of giving heat loss in different areas. The great disadvantage was the compressibility of the layer, although the development of pliable, permeable plastic materials allowed this difficulty to be overcome.

Hatfield (8) has described a unique heat flowmeter which consists of a thin disk (12 mm in diameter, 1.5 mm thick) of an alloy of tellurium coated on its two sides with fine copper gauze (160 meshes to the inch). Whyte and Reader (13) have reported on thermal gradient methods using thermocouples; Young and others (15) have developed thermal area meters combining the thermal gradient principle with thermal resistance thermometers.

INDIRECT CALORIMETRY

In indirect calorimetry the heat output of the subject is calculated from his oxygen consumption and his carbon dioxide output. Carlson, in a discussion of respiratory exchange (3), presented Weir's paper on the subject (12). The quantity of heat produced can be calculated from one of the following equations:

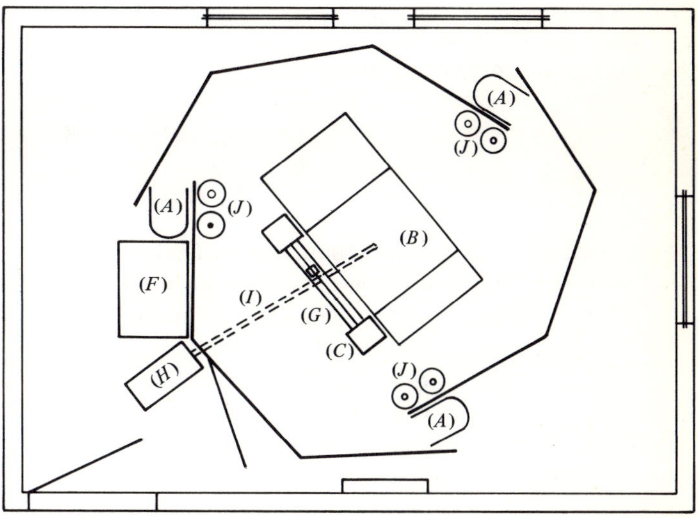

FIGURE A-4. Schematic diagram of partition calorimeter at John B. Pierce Foundation Laboratory. A: 2500 W heater; B: chair; C: aspiratory psychometer; F: chart table; G: platform scales; H: metabolism apparatus; I: hose connection; J: fans directed to floor. Aspiratory psychometer, Moll thermopile, and copper hemisphere radiometer complete equipment. (From C. E. A. Winslow, L. P. Herrington, and A. P. Gagge. A new method of partitional calorimetry. *Amer. J. Physiol.* **116:** 641–655, 1936.)

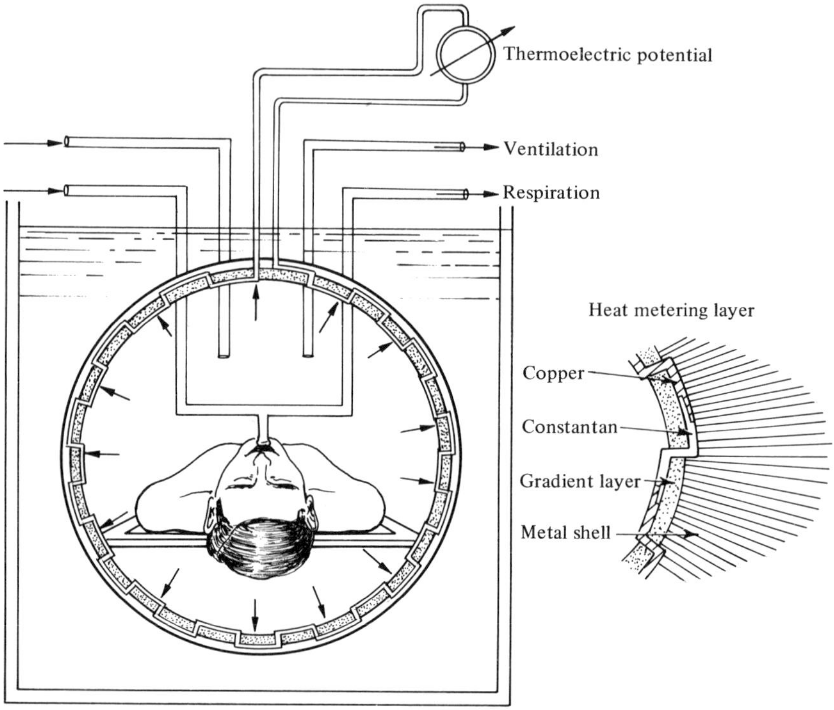

FIGURE A-5. Scheme and principle of the Benzinger gradient calorimeter. (From T. H. Benzinger and C. Kitzinger. Direct calorimetry by means of the gradient principle. *Rev. Sci. Instrum.* **20:** 849–860, 1949.)

$$Q = V_{O_2}\,(3.9 + 1.1R) \tag{1}$$
$$Q = 3.9\ V_{O_2} + 1.1\ V_{CO_2} \tag{2}$$
$$Q = V_E\,(1.046 - 5F_{EO_2}) \tag{3}$$

where Q = heat production in kcal,
V_{O_2} = oxygen consumption in 1,
V_{CO_2} = carbon dioxide production in 1,
V_E = volume of expired air in 1,
F_{EO_2} = fraction of oxygen in expired air, and
R = respiratory quotient.

Metabolic heat production can also be estimated from the CO_2 production:

$$Q = V_{CO_2}\,(3.9\frac{1}{R} + 1.1) \tag{4}$$

Assuming $R = 0.82$ and CO_2 in inspired air is zero:

$$Q = V_E \cdot F_{ECO_2} \cdot 5.85 \tag{5}$$

Spoor (10) has devised a method whereby a constant volume of air, exceeding in amount the maximum respiratory volume, is drawn past the subject. Enrichment of this constant volume with CO_2 from expired air will be proportional to metabolic rate. F_{ECO_2} will then vary directly with heat production (assuming R to be constant). The same principle can be applied to the estimation of Q from F_{EO_2} in equation (3).

The equations assume that the fraction of oxygen in the inspired air is 0.2093. For respiratory quotients between 0.718 to 1.0, the error in equations (3) and (5) is less than 1 in 600. A correction, assuming 10 to 15 per cent of the total calories arises from protein metabolism, is also included in the equations. With the greatest protein variation, this approximation will be in error by ±1 per cent. Therefore, all measurements of volume and gas concentrations should be this accurate. Thus, expired minute volume of 10 1/min should be accurate to 0.1 1 and expired oxygen fraction to 0.002. Calculations are made with all volumes and fractions at STPD, so the temperature, barometric pressure, and vapor pressure of the expired air should be determined. If the gas is collected below body temperature, it is usually assumed to be saturated.

MEASUREMENT OF VOLUME A Douglas bag and a suitably accurate gas meter or a spirometer are the traditional means of collecting and measuring the volume of expired air. Because of their limited capacity, they can be used for only a few minutes if the subject is performing strenuous exercise. (The Tissot gasometer has a capacity of 100 to 250 l.) Recent advances have led to the development of respirometers which are small, accurately built gas meters. They measure directly the volume of expired air and at the same time divert a small percentage of the air into a rubber bag for subsequent analysis. The volume flow per minute can also be measured by flow meters based on the pressure generated by air flow.

ANALYSIS OF EXPIRED AIR. Chemical methods are still the most accurate and are used to calibrate instruments that monitor changes in F_{EO_2} and F_{ECO_2}. The principle of the methods is the absorption of CO_2 with KOH and O_2 with a strong reducing solution from a known volume of air. The remainder of the gas is considered to be all N_2. The Haldane apparatus is suitable for gas samples of 5 to 10 ml and the Scholander micrometer gas analyzer will handle gas volumes of 0.5 ml with an accuracy of ±.015 volume per cent. (For details of the methods, see *Physiological Measurements of Metabolic Functions in Man,* by C. F. Consolazio, R. E. Johnson, and L. J. Pecora [6].)

Instruments which give a continuous recording of F_{EO_2} and F_{ECO_2} have been developed. The Pauling-Beckman oxygen analyzer determines the oxygen partial pressure of a sample by measuring its magnetic susceptibility. Oxygen is strongly paramagnetic, whereas other common gases are weakly

TABLE A-11

The Fraction of Fat in the Body Estimated from Specific Gravity or Body Water*

	Fraction Body Water	Fraction of Fat in Body					
Sp. Gr.†	$\left(4.340 - \frac{3.983}{\text{Sp. Gr.}}\right)$‡	$\left(\frac{5.548}{\text{Sp. Gr.}} - 5.044\right)$§	$\left(\frac{5.135}{\text{Sp. Gr.}} - 4.694\right)$‖	$\left(\frac{5.120}{\text{Sp. Gr.}} - 4.684\right)$#	$\left(\frac{4.235}{\text{Sp. Gr.}} - 3.827\right)$**	$\left(1 - \frac{\text{F. Body Water}}{0.718}\right)$††	$\left(1 - \frac{\text{F. Body Water}}{0.732}\right)$‡‡
1.090	0.686	0.046	0.017	0.013	0.058	0.045	0.068
1.085	0.669	0.069	0.039	0.035	0.076	0.068	0.086
1.080	0.652	0.093	0.061	0.057	0.094	0.092	0.109
1.075	0.635	0.117	0.083	0.079	0.112	0.116	0.132
1.070	0.618	0.141	0.105	0.101	0.131	0.139	0.156
1.065	0.600	0.165	0.128	0.124	0.149	0.164	0.180
1.060	0.582	0.190	0.150	0.146	0.168	0.189	0.205
1.055	0.565	0.215	0.173	0.169	0.187	0.213	0.228
1.050	0.547	0.240	0.196	0.192	0.206	0.238	0.253
1.045	0.528	0.265	0.220	0.216	0.226	0.265	0.279
1.040	0.510	0.291	0.244	0.239	0.245	0.290	0.303
1.035	0.492	0.316	0.267	0.263	0.265	0.315	0.328
1.030	0.473	0.342	0.291	0.287	0.285	0.341	0.354
1.025	0.454	0.369	0.316	0.311	0.305	0.368	0.380
1.020	0.435	0.395	0.340	0.336	0.325	0.394	0.406
1.015	0.416	0.422	0.365	0.360	0.345	0.421	0.432
1.010	0.396	0.449	0.390	0.385	0.366	0.448	0.459
1.005	0.377	0.476	0.415	0.410	0.387	0.475	0.485
1.000	0.357	0.504	0.441	0.436	0.408	0.508	0.512

*Lean body mass = Body weight (1 − Fraction of fat)

†Specific gravity can be estimated by water displacement, weighing underwater or from skinfold measurements: Sp. Gr. = 1.0909 − (0.0101*A* + 0.0090*B*): *A* = triceps fold in cm; *B* = subscapular fold in cm. *Source:* L. R. Pascale, M. I. Grossman, H. S. Sloane, and T. Frankel. Correlations between thickness of skinfolds and body density in 88 soldiers. *Human Biol.* **28:** 165–176, 1956.

‡Fraction of body water can be estimated by chemical analysis or from the empirical equation. *Source:* E. F. Osserman, G. C. Pitts, W. C. Welham, and A. R. Behnke. In vivo measurement of body fat and body water in a group of normal subjects. *J. Appl. Physiol.* **2:** 633–639, 1950.

§E. N. Rathbun and N. Pace. Studies in body composition. I. Determination of body fat by means of the specific gravity. *J. Biol. Chem.* **158:** 667–676, 1945.

‖A. R. Behnke, E. F. Osserman, and W. E. Welham. Lean body mass. *A.M.A. Arch. Int. Med.* **91:** 585–601, 1953.

#A. Keys and J. Brozek. Body fat in adult man. *Physiol. Rev.* **33:** 245–326, 1953.

**T. M. Fraser. Biomedical techniques for use in manned space laboratories. Part III. Anthropometry and body composition. Rep. IX. Contract No. NASr-115, NASA, 1966.

††Same source as ‡. ‡‡Same source as #.

TABLE A-12
Surface Area (m^2) from Height (cm) and Weight (kg)*

W(kg) \ H(cm)	50	155	160	165	170	175	180	185	190	195
40	1.30	1.33	1.36	1.39	1.42	1.45				
45	1.37	1.40	1.43	1.46	1.49	1.52				
50	1.44	1.47	1.50	1.53	1.56	1.59				
55	1.49	1.52	1.55	1.58	1.61	1.65	1.69	1.72	1.76	
60				1.65	1.68	1.72	1.76	1.79	1.83	
65				1.70	1.74	1.78	1.82	1.85	1.89	
70				1.77	1.80	1.84	1.88	1.92	1.96	
75				1.82	1.86	1.89	1.94	1.98	2.02	2.06
80					1.92	1.96	2.00	2.05	2.09	2.14
85					1.96	2.01	2.05	2.10	2.14	2.19
90					2.02	2.07	2.11	2.16	2.20	2.25
95					2.07	2.12	2.16	2.20	2.25	2.30
100					2.12	2.16	2.21	2.26	2.30	2.35

*Calculated from the equation: $m^2 = W^{0.425} \cdot H^{0.725} \cdot 0.007184$. Interpolate for values not shown.

TABLE A-13
Conversion Table of kg to $kg^{3/4}$

kg		0.01	0.02	0.03	0.04	0.05	0.06	0.07	0.08	0.09
		0.0316	0.0532	0.0720	0.0894	0.106	0.121	0.136	0.150	0.164
0.1	0.177	0.191	0.204	0.217	0.299	0.241	0.253	0.265	0.276	0.288
0.2	0.299	0.313	0.323	0.333	0.343	0.353	0.364	0.375	0.385	0.395
0.3	0.405	0.415	0.425	0.435	0.445	0.455	0.465	0.475	0.484	0.493
0.4	0.503	0.513	0.522	0.531	0.540	0.549	0.558	0.567	0.576	0.585
0.5	0.594	0.603	0.612	0.621	0.530	0.638	0.647	0.656	0.665	0.674
0.6	0.682	0.690	0.699	0.707	0.716	0.724	0.733	0.741	0.749	0.757
0.7	0.765	0.774	0.782	0.790	0.798	0.806	0.814	0.822	0.830	0.838
0.8	0.846	0.854	0.862	0.870	0.878	0.885	0.893	0.900	0.908	0.916
0.9	0.923	0.931	0.939	0.946	0.954	0.962	0.970	0.978	0.986	0.993

kg		1	2	3	4	5	6	7	8	9
		1.00	1.68	2.28	2.83	3.34	3.83	4.30	4.75	5.20
10	5.62	6.04	6.44	6.85	7.24	7.62	8.00	8.38	8.74	9.10
20	9.45	9.81	10.2	10.5	10.8	11.2	11.5	11.8	12.2	12.5
30	12.8	13.1	13.5	13.8	14.1	14.4	14.7	15.0	15.3	15.6
40	15.9	16.2	16.6	16.8	17.1	17.4	17.7	17.9	18.2	18.5
50	18.8	19.1	19.4	19.6	19.9	20.2	20.5	20.8	21.0	21.3
60	21.6	21.9	22.1	22.4	22.6	22.9	23.2	23.5	23.7	23.9
70	24.2	24.5	24.7	25.0	25.2	25.5	25.7	25.9	26.2	26.5
80	26.7	27.0	27.2	27.5	27.7	28.0	28.3	28.5	28.8	29.0
90	29.2	29.5	29.7	29.9	30.1	30.4	30.6	30.9	31.2	31.4
100	31.6	31.9	32.1	32.4	32.6	32.8	33.1	33.3	33.5	33.7
110	34.0	34.2	34.4	34.7	34.9	35.1	35.4	35.6	35.8	36.0

diamagnetic. The magnetic susceptibility of oxygen can be regarded as a measure of the tendency of an oxygen molecule to become temporarily magnetized when placed in a magnetic field. The susceptibility measurement is made in an analysis cell where a test body is suspended in a nonuniform field. Any change in partial pressure causes the test body to rotate. The magnetic forces are measured by applying an electrostatic force equal and opposite to the magnetic force.

Carbon dioxide analyzers are based on the selective absorption by the gas of radiation in the infrared region of the spectrum (CO 4.7; H_2O 6.0; CO_2 4.3 micron wavelength absorption bands). The instrument consists of a detecting unit which generates an electrical signal proportional to the concentration and an amplifier which translates the signal into a reading on a panel meter and into an output which can be connected to a recorder.

In polarographic transducers, current flow between two electrodes is a function of the amount of oxygen in the liquid or gas environment being measured. The reference electrode is made of silver or silver-silver chloride alloy. The active electrode generally is made of platinum. The electrolyte usually is potassium chloride or sodium chloride. A small voltage (0.5 to 0.7 V) is impressed across the electrodes, causing a small current flow. This current flow decreases as the cell becomes polarized by the concentration of hydrogen ions at the platinum cathode. The electrodes and electrolyte are sealed behind a semipermeable membrane which prevents contamination of the electrolyte but permits oxygen to pass through. The oxygen molecules are drawn to the platinum electrode and combine with the hydrogen ions. This depolarizes the cell and there is an increase in current flow. The amount of current flow is proportional to the oxygen concentration of the gas.

References

1. Benedict, F. G., and T. M. Carpenter. Respiration calorimeters for studying the respiration exchange and energy transformations of man. Publ. No. 123. Washington, D.C.: Carnegie Institute of Washington, 1910.
2. Benzinger, T. H., and C. Kitzinger. Direct calorimetry by means of the gradient principle. *Rev. Sci. Instrum.* **20:** 849–860, 1949.
3. Carlson, L. D. Respiratory exchange. In: *Methods in Medical Research.* Chicago: Year Book Medical Publishers, 1954, vol. 6, pp. 60–73.
4. Carlson, L. D., N. Honda, T. Sasaki, and W. V. Judy. A human calorimeter. *Proc. Soc. Exp. Biol. Med.* **117:** 327–331, 1964.
5. Carrol, D. P., and J. Visser. Direct measurement of convective heat loss from human subjects, *Rev. Sci. Instrum.* **37:** 1174–1180, 1966.
6. Consolazio, C. F., R. E. Johnson, and L. J. Pecora. *Physiological Measurements of Metabolic Functions in Man.* New York: McGraw-Hill, 1963.
7. Day, R., and J. D. Hardy. Respiratory metabolism in infancy and childhood.

XXVI. A calorimeter for measuring the heat loss of premature infants. *Amer. J. Dis. Child.* **63:** 1086–1095, 1942.
8. Hatfield, H. S. A heat-flow meter. *J. Physiol.* **111:** 10p–11p, 1949.
9. Murlin, J. R., and A. C. Burton. Human calorimetry. I. A semiautomatic respiration calorimeter. *J. Nutr.* **9:** 233–260, 1935.
10. Spoor, H. J. Application of the infrared analyzer to the study of human energy metabolism. *J. Appl. Physiol.* **1:** 369–384, 1948.
11. Visser, J. and T. Hodgson. The design of a human calorimeter for the determination of body heat storage. *South African Engineer* **9:** 243–269, 1960.
12. Weir, J. B. de V. New methods for calculating metabolic rate with special reference to protein metabolism. *J. Physiol.* **109:** 1–9, 1949.
13. Whyte, H. M., and S. R. Reader. Needle thermocouples. *J. Appl. Physiol.* **7:** 623–627, 1952.
14. Winslow, C. E. A., L. P. Herrington, and A. P. Gagge. A new method of partitional calorimetry. *Amer. J. Physiol.* **116:** 641–655, 1936.
15. Young, A. C., H. L. Burns, W. E. Quinton, and L. D. Carlson. Temperature gradient calorimetry. *USAF Tech. Rep. No. 6248,* 1951.

Additional Reading

Hardy, J. D. Temperature measurement and calorimetry. In: *Temperature: Its Measurement and Control in Science and Industry.* Part 3, J. D. Hardy, ed. New York: Reinhold, 1963.

Kleiber, M. Calorimetric measurements. In: *Biophysical Research Methods.* Edited by F. M. Uber. New York: Interscience, 1950, Chap. VI.

APPENDIX IV

Models of Energy Exchange

The Woodcock-Thwaites-Breckenridge Model

Figure A-6 shows a complete circuit for an idealized man-clothing-environment system. The elements have been incorporated into an electrical analogue.

REFERENCE

Woodcock, A. H., H. L. Thwaites, and J. R. Breckenridge. An electrical analogue for studying heat transfer in dynamic situations. Hdqtrs. Quartermaster Research and Engineering Command, TR-EP-86, 1958.

The Wissler Model—Six-Element Man

Figure A-7 depicts the Wissler model, which includes the distribution of heat generation; the conduction of heat in tissue; the convection of heat by flowing blood; the loss of heat by radiation, convection, and evaporation at the surface; the loss of heat through the respiratory tract; and the countercurrent heat exchange between large arteries and veins. The numbers show the six elements considered separately. T is temperature of a(arterial) or v(venous) blood, and UA is heat exchange.

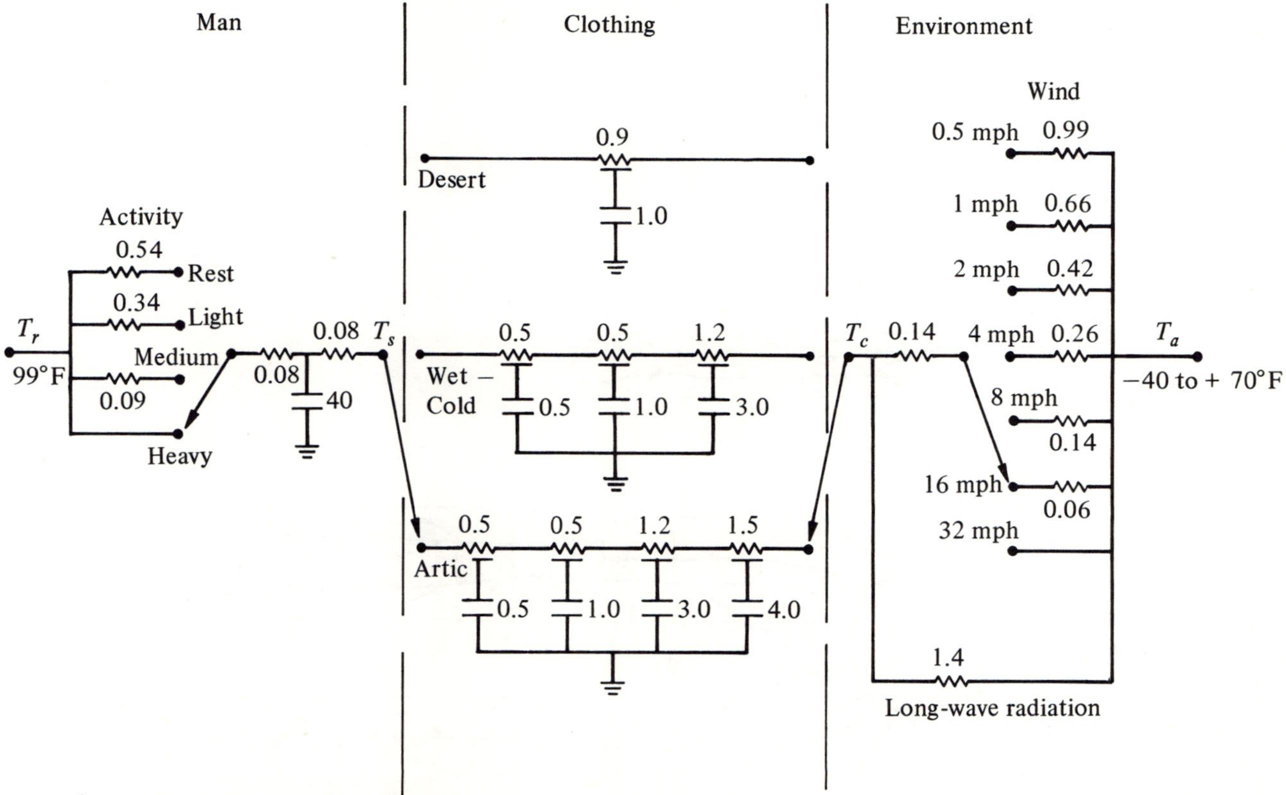

⏚ Denotes constant reference level.

Note: All thermal resistances in clo
All thermal capacities in Btu/°F

FIGURE A-6. Complete thermal circuit for the idealized man-clothing-environment system.

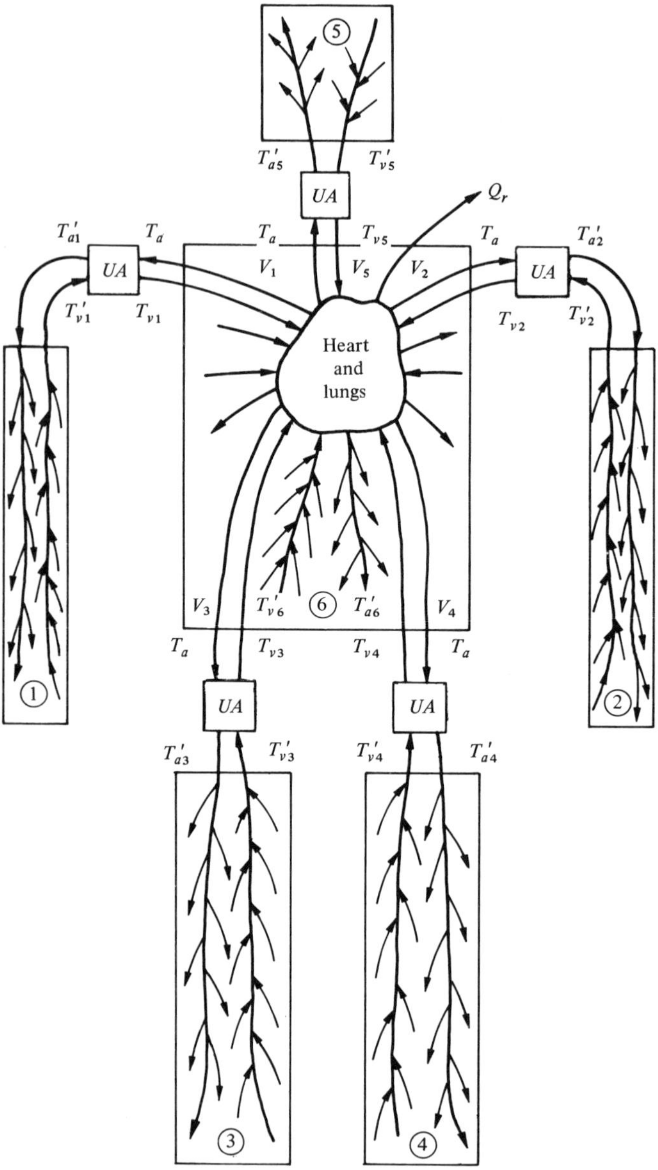

FIGURE A-7. The six-element man. (From E. H. Wissler. Steady state temperature distribution in the human. *University of Texas, Mathematical Studies in Thermal Physiology, Report No. 1,* 1959.)

REFERENCES

Wissler, E. H. Steady state temperature distribution in the human. *University of Texas, Mathematical Studies in Thermal Physiology, Report No. 1,* 1959.

Wissler, E. H. Transient-state temperature distribution in the human. *University of Texas, Mathematical Studies in Thermal Physiology, Report No. 2,* 1959.

Wissler, E. H. An analysis of factors affecting temperature levels in the nude human. *University of Texas, Mathematical Studies in Thermal Physiology, Report No. 4,* 1961.

The Hammel Model

The organization of Figure A-8 is unusual in that it illustrates a format often used by systems engineers for a feedback control system. Such a block diagram indicates the direction of flow of information and the transformations occurring along the way. Because this system has at least one feedback loop as well as a constant command, it is said to be a closed-loop feedback regulator, where core temperature is probably the directly controlled variable. Each block in the diagram represents a transformation, and the law of the transformation—that is, the input-output relationship—determines the output for any given input. Each circle represents a mixing point where quantities of the same quality and same dimension are combined. The mixing points in the controlling system probably combine nerve impulses and therefore are shown to be adders or subtractors depicting facilitation or inhibition of nervous activity. The outputs of the controllers are said to be the output-forcing functions which, along with the disturbances, become the inputs to the controlled system. With further transformation within the controlled system by organs of the body, each of the forcing functions becomes a rate of heat production or heat loss. The outputs of the controlled system are the body temperatures which are transduced and fed back in some way to the controlling system as nervous activity. Hammel has suggested the following relationship between inputs and outputs:

$$R - R_o = \alpha_R \quad (T_{hypo} - T_R) \pm \Delta T_R - \beta(40 - T_s) - \beta T_s + \gamma(T_s - 37) + \gamma T_s + \delta T_b$$

where $R - R_o$ is the thermoregulatory response (metabolism, vasomotor activity, sweating, behavior, and so on); R_o is the basal level when $T_{hypo} = T_R$; T_{hypo} = the hypothalamic temperature is the feedback signal; T_R = the set point for response; R is the reference input signal; T_s and T_b = skin and deep body temperature.

REFERENCE

Hammel, H. T. Regulation of internal body temperature. *Annu. Rev. Physiol.* **30:** 641–710, 1968.

The Brown Model

A symbolic diagram of the complete circuit used for the analog computation is shown in Figure A-9. In this figure the four thermal elements of the

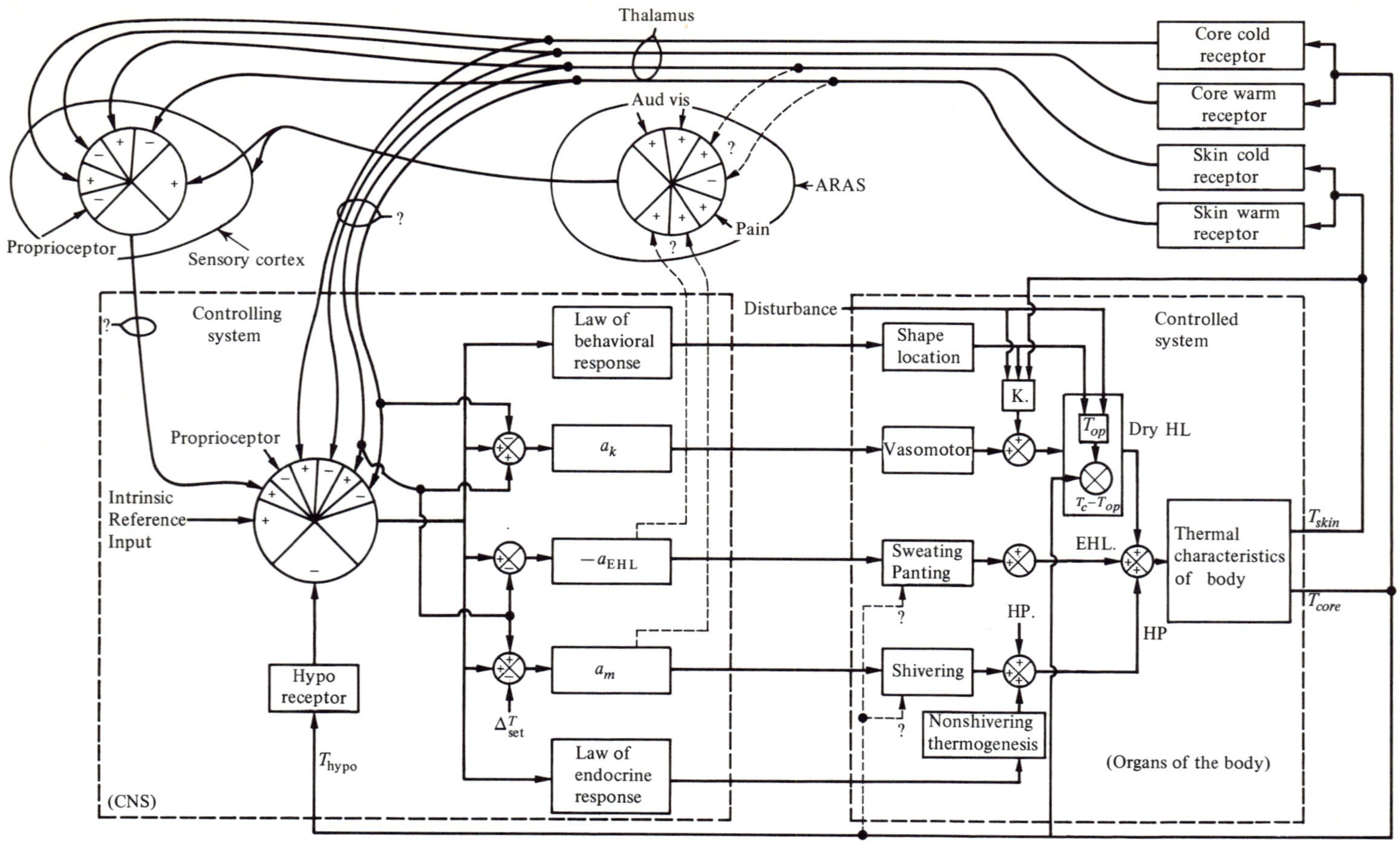

FIGURE A-8. A schema for the controlling and controlled systems for the regulation of internal body temperature. ARAS is the ascending reticular activating system.

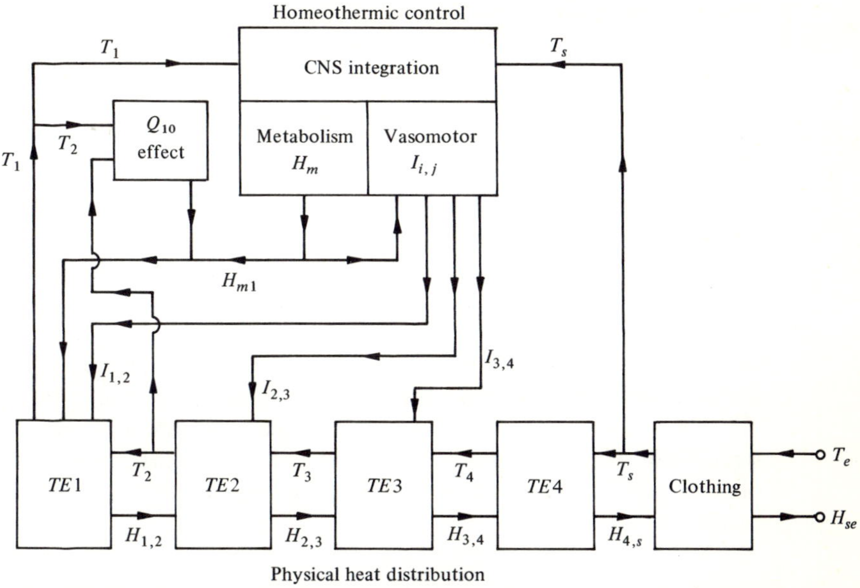

FIGURE A-9. A diagram of the complete circuit used for the analog situation.

body (*TE* 1 to 4) and the element representing the clothing are arranged in series. Given the temperature of the adjacent peripheral element and the heat influx from the adjacent central one, each thermal element computes its mean temperature and peripheral heat efflux. To perform this computation, the rate of internal generation of metabolic heat, the heat capacity, and the equivalent tissue conductivity must be known. The heat capacity is a constant determined by the anatomy of the subject being simulated. The metabolic rate and equivalent tissue conductivity are determined in part by central temperature (T_1) and surface temperature (T_s) through the mechanism of central nervous system integration, and in part by the direct effects of temperature upon metabolic rate (Q_{10} effect) and the direct influence of H_m upon $I_{1,2}$.

All the dependent variables are computed by the circuit with the exception of the programmed variable representing the temperature of the environment. This corresponds to the fact that for water immersion the water temperature is the sole measurement of environmental stress in this analog; the other dependent variables represent physical or physiologic response to this stress and thus may be computed.

REFERENCES

Brown, A. C. Analog computer simulation of temperature regulation in man. AF TR#AMRL-TDR-63-116, Wright-Patterson Air Force Base, Ohio, 1963.

Brown, A. C. Further development of the biothermal analog computer. AF TR#-AMRL-TR-66-197, Wright-Patterson Air Force Base, Ohio, 1966.

The Stolwijk-Hardy Model

In the Stolwijk-Hardy model (Figure A-10) the human body is represented by three cylinders (the head, the trunk, and the extremities) divided into eight compartments. All cylinders exchange heat by convection with a central blood compartment and with skin layers by conduction. The dimensions of the model have been calculated from anatomical data for a 71-kg "standard man." The authors developed eight heat-flow equations for the compartments which are suitable for programming on an analog computer. These equations incorporate heat input (M_o = basal, ΔM = shivering, E = exercise), heat loss or gain (T_A = temperature of environment, E_V = evaporation), and blood flow (MBF = muscle blood flow; SBF = skin blood flow). The passive model was found to be a good representation of the thermal exchanges in man.

The authors have also shown how this passive system can be coupled with a regulator which accepts signals and which exerts corrective effector actions on the passive system if the signals indicate a deviation from preferred conditions.

REFERENCE

Stolwijk, J. A. J. and J. D. Hardy. Temperature regulation in man—A theoretical study. *Pflueger. Arch. Ges. Physiol.* **291:** 129–162, 1966.

The Aschoff-Wever Extremity Model

The analogue depicted in Figure A-11 was constructed to simulate the heat loss and heat exchanging in the extremity.

REFERENCES

Aschoff, J. Wärmeaustausch in einer modellextremität. *Pflueger. Arch. Ges. Physiol.* **264:** 260–271, 1957.

Aschoff, J., and R. Wever. Modellversuche zum gegenstrom-wärmeaustausch in der extremitat. *Z. Ges. Exp. Med.* **130:** 385–395, 1958.

Aschoff, J., and R. Wever. Wirkungen des wärme-kurzschlusses in einer modell-extremität. *Pflueger. Arch. Ges. Physiol.* **267:** 120–127, 1958.

The Hsia-Hsu-Webster-Carlson Model of Periphery

This analog contains an assemblage of modules, each of which contains the circuit elements needed for simulation of heat production, thermal inertia, heat flow, and blood flow. Figure A-12 is a simplified block diagram describing the interrelationships between various variables and parameters. The circula-

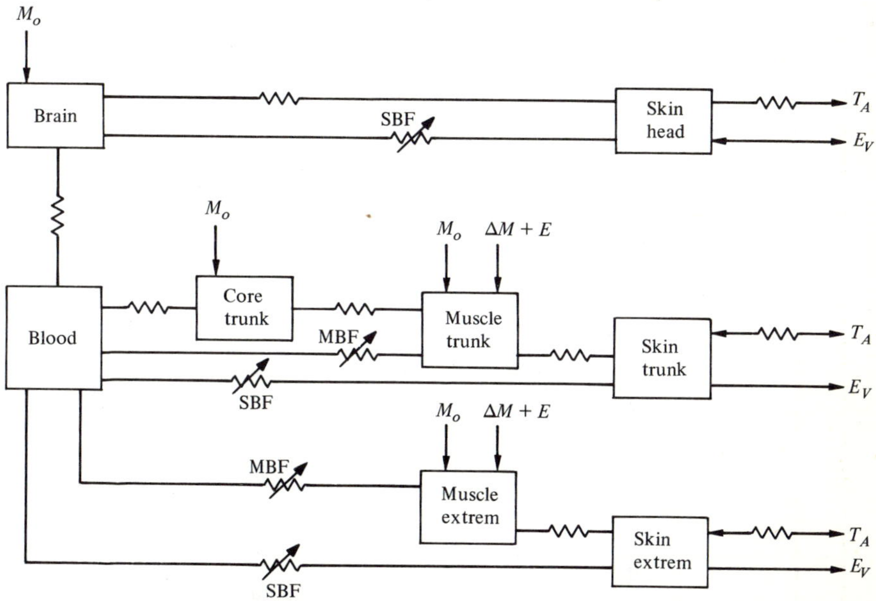

FIGURE A-10. Block diagram of relationship between the various compartments of the thermal model of man.

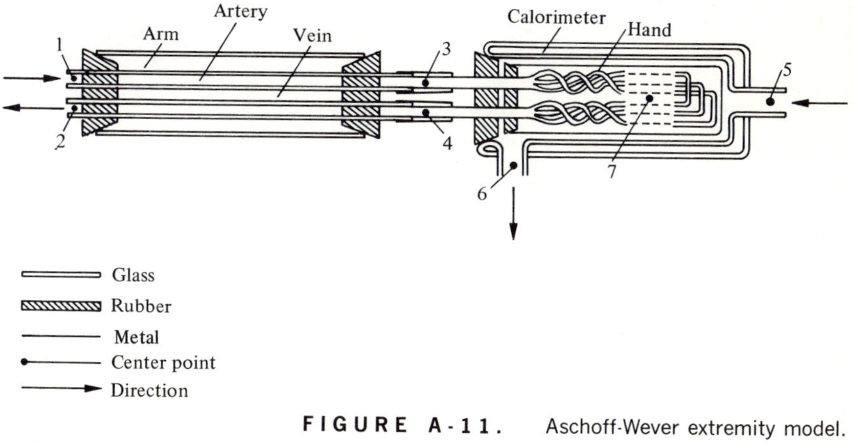

FIGURE A-11. Aschoff-Wever extremity model.

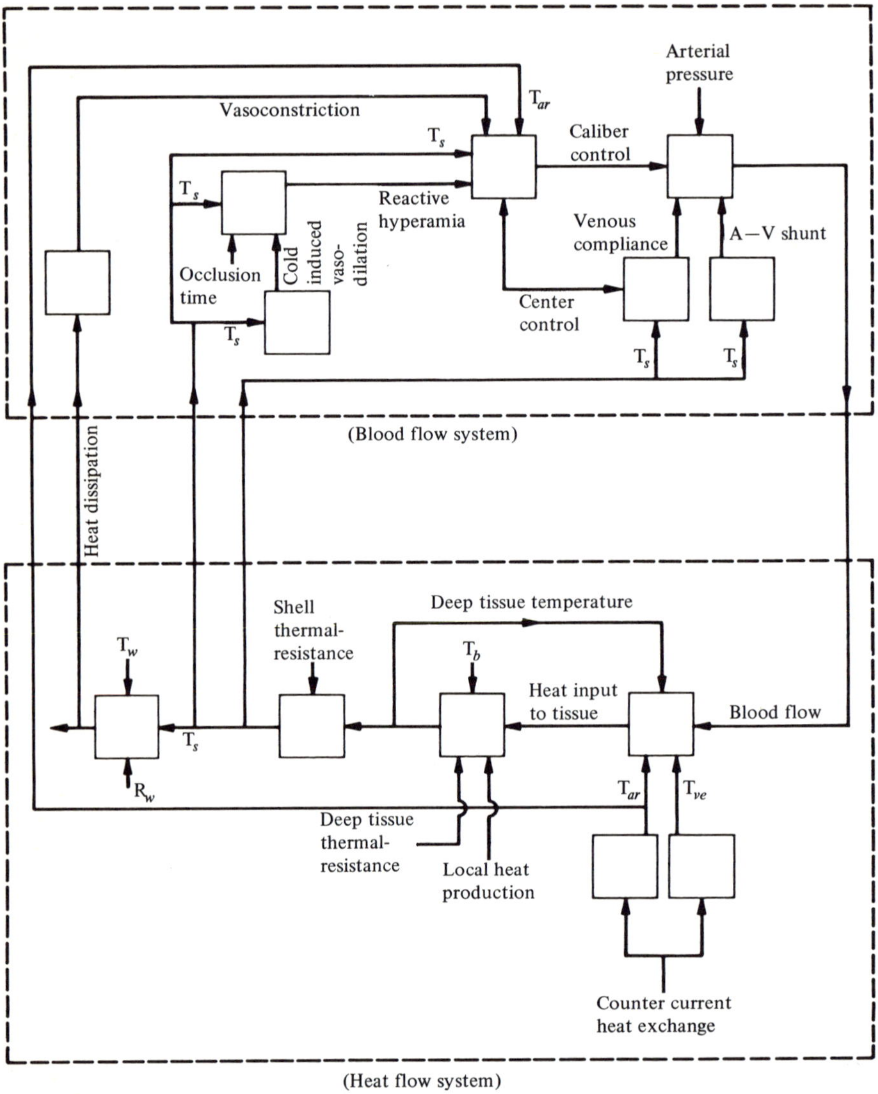

FIGURE A-12. Hsia-Hsu-Webster-Carlson model of periphery.

tory system controls the blood flow which serves as a major heat source in the thermal system; on the other hand, variations in temperature and heat content of the thermal system are constantly fed back to regulate the blood flow.

REFERENCE

Hsia, T. C., T. G. Hsu, M. E. D. Webster and L. D. Carlson. A heat transfer model of the human limb. (Unpublished).

Index